Electronic De

F R Connor

Ph D, M Sc, B Sc (Eng) Hons, ACGI,
C Eng, MIEE, MIERE, M Inst P

Edward Arnold
A division of Hodder & Stoughton
LONDON NEW YORK MELBOURNE AUCKLAND

First published in Great Britain 1980
Reprinted with corrections 1984, 1989

British Library Cataloguing in Publication Data

Connor, Frank Robert
Electronic devices.
1. Electronic apparatus and appliances
I. Title
621.381 TK7870

ISBN 0–7131–3413–5

Photoset in India by The Macmillan Co of India Ltd, Bangalore
Printed and bound in Great Britain for Edward Arnold, the educational, academic and medical publishing division of Hodder and Stoughton Limited, 41 Bedford Square, London WC1B 3DQ by
J. W. Arrowsmith Ltd, Bristol

Preface

This is an introductory book on the important topic of *Electronic Devices*. Considerable development work has taken place in this subject during recent years, especially in the field of solid-state devices. However, the book endeavours to present the basic ideas of both vacuum and solid-state devices in a concise and coherent manner. Moreover, to assist in the assimilation of these basic ideas, many worked examples from past examination papers are provided to illustrate clearly the application of the fundamental theory.

The first part of the book is devoted to some fundamental atomic and semiconductor theory which provides an understanding of the physical principles underlying many electronic devices. The following chapter is devoted to the important area of solid-state devices and emphasis is placed on both the principles and applications of these devices. In a subsequent chapter, the treatment of vacuum devices is intended to cover as wide an area as possible, since a knowledge of these devices is still essential for many applications. The book ends with a useful set of appendices to provide a more theoretical approach to some of the devices and is intended for the reader who wishes to pursue the matter further. Also included are brief details of some more recent devices such as the Impatt diode and charge-coupled device (CCD).

This book will be found useful by students preparing for University examinations, degrees of the Council of National Academic Awards, examinations of the Council of Engineering Institutions and for other qualifications such as Higher National Certificates and Higher National Diploma. It will also be of considerable use to practising scientists and engineers in Industry who require a ready source of basic knowledge to help them in their applied work.

Acknowledgements

The author sincerely wishes to thank the Senate of the University of London and the Council of Engineering Institutions for permission to include questions from past examination papers. The solutions and answers provided are his own and he accepts full responsibility for them.

Grateful thanks are also due to RCA Solid State Division, Somerville, New Jersey and TRW Semiconductors, Lawndale, California for their kindness in providing the photographs shown in Fig. 4 and to Texas Instruments Ltd.,

Manton Lane, Bedford for kindly providing the chip photograph of the TMS9900 shown in Fig. 82.

Finally, the author would like to thank the publishers for many useful suggestions and will be grateful to his readers for drawing his attention to any errors which may have occurred.

1979 FRC

Contents

Symbols

c	velocity of light
e	electronic charge
f	frequency
f_c	cut-off frequency
f_p	pump frequency
f_s	signal frequency
g_m	mutual conductance
h	Planck's constant
$\hbar$	$h/2\pi$
k	Boltzmann's constant
	wave number
m	electron mass
m_0	rest mass (electron)
m_e	effective mass (electron)
m_h	effective mass (hole)
m_p	proton mass
n	carrier density (electrons)
	n-type material
n_i	intrinsic carrier density
p	carrier density (holes)
	p-type material
q	electric charge
r	radius
r_a	anode resistance
r_d	drain resistance
t	time
v	velocity
v_n	drift velocity (electrons)
v_p	drift velocity (holes)
w	width
A	any constant
B	magnetic flux density
D	charge displacement density
D_n	diffusion coefficient (electrons)
D_p	diffusion coefficient (holes)
E	electric field intensity
E_0	rest mass energy (electron)
I_B	base current
I_C	collector current
I_{CBO}	collector-base leakage current
I_{CEO}	collector-emitter leakage current
I_D	drain current

I_{DS}	saturation drain current
I_E	emitter current
J	current density
J_{en}	current density of electrons (*n*-type material)
J_{ep}	current density of electrons (*p*-type material)
J_{hn}	current density of holes (*n*-type material)
J_{hp}	current density of holes (*p*-type material)
K	Kelvin
L_e	diffusion length (electrons)
L_h	diffusion length (holes)
N_a	density of acceptor atoms
N_d	density of donor atoms
R_H	Hall coefficient
T	absolute temperature
	kinetic energy
V	potential energy
	voltage
V_{CB}	collector-base voltage
V_{CE}	collector-emitter voltage
V_{DS}	drain-source voltage
V_{GS}	gate-source voltage
V_p	pinch-off voltage
W	energy
W_c	energy (conduction band edge)
W_f	work function
W_g	energy (band gap)
W_v	energy (valence band edge)
W_B	surface barrier energy
W_F	equilibrium Fermi level
Z	atomic number
α	common-base current factor
β	common-emitter current factor
γ	emitter efficiency
ε_0	permittivity of free space
ε	permittivity of medium
λ	wavelength
μ_n	mobility (electrons)
μ_0	permeability of free space
μ_p	mobility (holes)
ρ	resistivity
σ	conductivity
τ_e	mean lifetime (electrons)
τ_h	mean lifetime (holes)
ϕ	work function

ψ	wave function
ω	angular frequency

Abbreviations

U.L.	University of London, B Sc(Eng) examination
C.E.I.	Council of Engineering Institutions, Part 2 examination

Physical constants

c	3×10^{8} m s^{-1}	m_p	$1{\cdot}67 \times 10^{-27}$ kg
e	$1{\cdot}6 \times 10^{-19}$ C	μ_0	$4\pi \times 10^{-7}$ H m^{-1}
h	$6{\cdot}63 \times 10^{-34}$ J-s	ε_0	$8{\cdot}85 \times 10^{-12}$ F m^{-1}
k	$1{\cdot}38 \times 10^{-23}$ J K^{-1}	1 eV	$1{\cdot}60 \times 10^{-19}$ J
m	$9{\cdot}11 \times 10^{-31}$ kg		

Physical properties

	Ge	Si
Atomic number	32	14
Atomic weight	72.6	28.08
μ_n at 300 K (m s^{-1}/V-m)	0.39	0.14
μ_p at 300 K (m s^{-1}/V-m)	0.19	0.05
D_n (room temperature) (m^2 s^{-1})	0.01	0.004
D_p (room temperature) (m^2 s^{-1})	0.005	0.001

1

Introduction

The subject of Electronics covers a wide field of knowledge which is difficult to define exactly. It springs from the generic term 'electron' and is largely connected with the study of electron motion. Hence, this book is mainly concerned with the study and use of vacuum and solid state devices of various kinds.

The development of electronic devices over the last hundred years has advanced considerably, and this has had a profound effect on industry and society. On the industrial side, it has led to a large electronics industry which is the key to development and prosperity in the industrialised nations of the world, notably Europe and America. On the social side, it has permeated almost all branches of human activity including education, commerce and social welfare. Moreover, it has led to world-wide repercussions, especially in the field of communications, and is now playing a vital role in the solution of present day problems, mainly through the application of computers in various fields of technology.

Historically, this development is the result of a series of important discoveries which date back to the end of the 19th century. The first of these took place in 1883 when Edison was investigating light emitted from a carbon filament lamp. During his attempts to increase the lifetime of the filaments, he observed that current flowed across the vacuum between the filament and a 'positive' plate inserted near it. This phenomenon, known as the *Edison effect*, led Fleming in 1904 to construct the diode, which is well-known for its rectifying property. This diode *valve* was used by Fleming as a detector of radio waves.

The addition of a grid to Fleming's diode led to the discovery of the *triode* by de Forest in 1906, and opened the way for the future development of *wireless* communication. This was initially made possible by the amplifying properties of the triode, and by its subsequent use in the generation of oscillations. It also led to the development of various vacuum devices for use in receiving and transmitting equipment, and to various gas tubes, notably the thyratron, for use in high power applications such as radar and industrial control.

However, the next major development was to take place in an entirely different field known later as solid-state electronics. The foundations of this new field were laid around 1900 by the discovery that certain materials, such as copper oxide, silicon and lead sulphide, have rectifying properties. The nature of this effect was not fully understood until many years later.

The work of Bardeen and Brattain[1] on semiconductors which have a conductivity between that of a metal and that of an insulator, depending on temperature, led to the development of the point-contact transistor in 1948 and to the development of the junction transistor by Shockley, Sparks and Teal[2] in 1951. The two types of transistors are illustrated in Fig. 1.

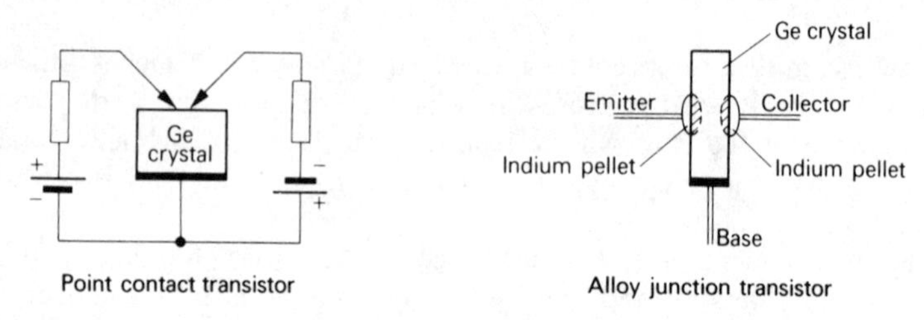

Fig. 1

Unlike the vacuum devices which owe their conduction to negatively charged carriers called *electrons*, the junction transistor depends upon the motion of negatively charged *electrons* and positively charged *holes*. It consists basically of a piece of semiconducting material such as germanium or silicon. A thin central region of given conductivity is sandwiched between two outer regions, both of which are of opposite conductivity to the centre. Such a *pnp* or *npn* device contains free electrons and holes, which are able to move through the crystal lattice structure under the application of a small electrical voltage.

The transistor therefore behaves like a triode, with the three electrodes known as emitter, base and collector. Such a device has considerable advantage over the triode because it is small in size, requires no heating power or vacuum, and only needs a low d.c. voltage. Consequently, it has largely replaced vacuum devices in many low-power applications and has given a tremendous boost to the electronics industry. Moreover, it has opened up a wide range of new solid-state devices and more recently, to the unipolar transistor or *field-effect* transistor (FET)[3] shown in Fig. 2.

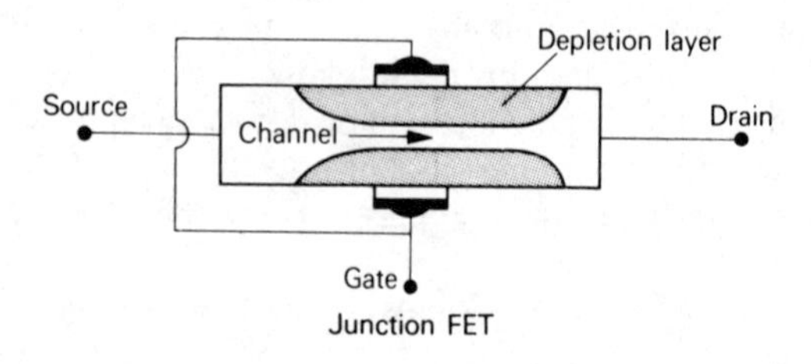

Junction FET

Fig. 2

In the FET, the flow of charge carriers between the *source* and *drain* is controlled by the voltage on the *gate*, which exerts a 'field-effect' on the channel. The gate voltage determines the width of the depletion layer, and hence the width of the channel which controls the current through the device. Two versions of the FET are the *junction-gate* FET (JFET), shown in Fig. 2, and the *insulated-gate* FET (IGFET) shown in Fig. 3. The insulated-gate FET is in effect like a pentode and has similar characteristics.

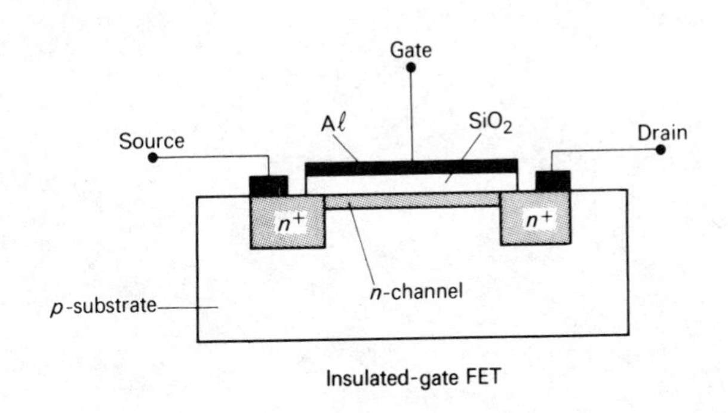

Fig. 3

The present state of the art has led to numerous applications of transistors in many electronic fields such as communications and computers. The latter has in turn prompted a further important development known as micro-electronics. The considerable improvement in materials and techniques in recent years has led to the construction of many elements such as transistors, resistors and capacitors, all on the same piece of material or *chip* which is known as an integrated circuit (IC)[4]. The two main types of IC are the *monolithic* integrated circuit and the *hybrid* integrated circuit which are exemplified in Fig. 4.

The rapid development of such silicon integrated circuits is due mainly to planar transistor technology. A diffusion process, together with lithographic techniques, is used to produce several circuits on a small silicon slice whose dimensions are about 1·5 mm × 1·5 mm. An ever increasing demand to put more circuits on a small chip has led even further to *large scale integration* (LSI), for the fabrication of sub-systems rather than circuits. A typical example is the microprocessor.[5]

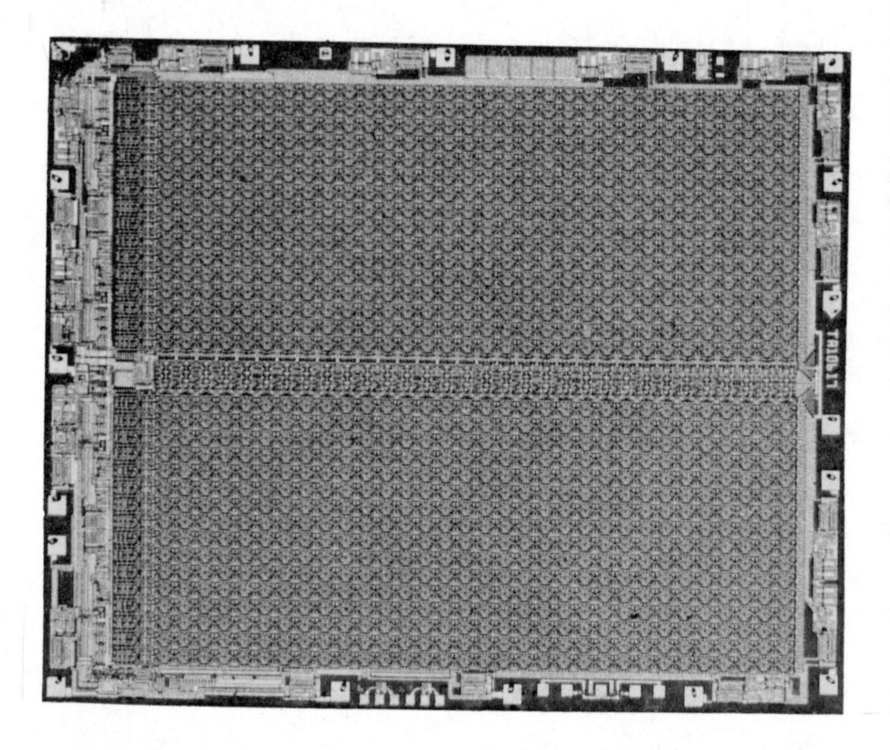

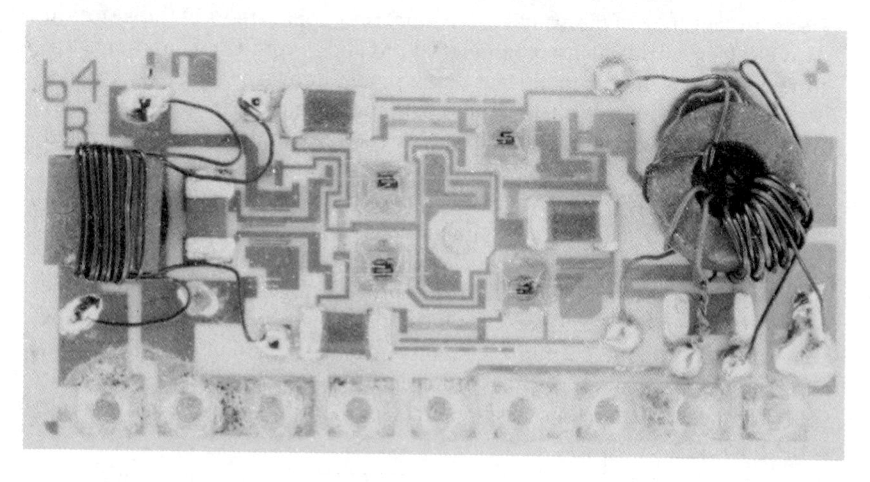

Fig. 4 (above) Monolithic 4K Random access memory. (By courtesy of RCA Solid State Division.)
(below) RF Linear Hybrid Amplifier. (By courtesy of TRW Semiconductors.)

2

Atomic theory[6]

Matter is composed of elements, and elements are made up of atoms. The atom is regarded as the smallest part of an element which retains the physical and chemical identity of that element. Up to the present time, 105 elements have been identified (see Appendix A) and the atoms of all these elements are made up of various numbers of the fundamental particles known as the proton, neutron and electron.

In 1911, Rutherford experimentally proved that an atom consists of a central, positively charged body called the *nucleus* around which negatively charged electrons revolve in circular orbits. It is now known that the nucleus is composed mainly of positively charged particles called *protons* and some neutral particles called *neutrons*. The atom as a whole is neutral because the positive charge Ze on the nucleus equals the negative charge $-Ze$ of the revolving electrons, where Z is the atomic number of the element and $-e$ is the negative charge on the electron. This is illustrated in Fig. 5 for the hydrogen atom.

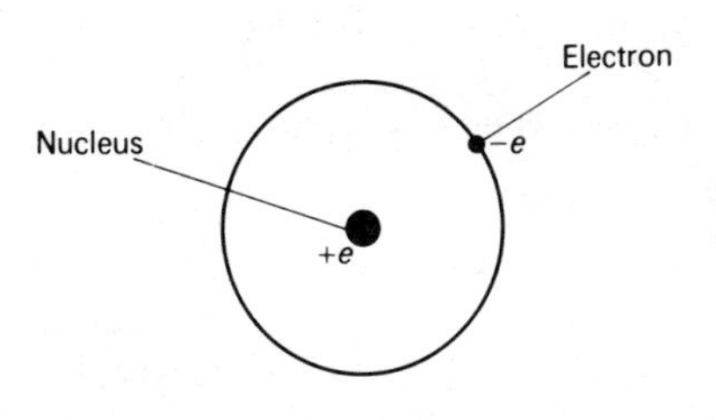

Fig. 5

2.1 Fundamental particles

Experimental evidence concerning the proton indicates that it has a mass of $1{\cdot}672 \times 10^{-27}$ kg, a positive charge of $1{\cdot}602 \times 10^{-19}$ coulomb and a diameter of about 6×10^{-15} m. The number of protons in the nucleus of an atom determines its atomic number Z and for the lightest element hydrogen, $Z = 1$ while for the heavy radioactive element uranium, $Z = 92$.

With the exception of hydrogen, neutrons are also present in the nucleus in about the same number as protons. The mass of a neutron is $1{\cdot}675 \times 10^{-27}$ kg which is about that of a proton, but it has no charge. The exact role of neutrons in the nucleus is not certain, but they appear to provide a binding force with protons, which keeps the nucleus together.

In contrast, the electron is the lightest known particle, with a mass of $9{\cdot}11 \times 10^{-31}$ kg and a negative charge of $1{\cdot}602 \times 10^{-19}$ coulomb. Its diameter is about $5{\cdot}6 \times 10^{-15}$ m and so it is about the same size as a proton or neutron. However, due to its low inertia, it plays a vital role in electronics and it can behave as a particle or a wave. This is known as the wave-particle duality, and consequently the 'size' of the electron is not well defined.

In addition to these particles, two other 'particles' which are necessary in understanding certain physical phenomena are the *photon* and the *hole*. The photon is generally regarded as a wave-packet or *quantum* of electromagnetic energy $W = hf$ where h is Planck's constant and f is the frequency of the radiation. The energy of light waves, for example, is conveyed by photons or quanta. The hole, however, appears to behave as if it were a particle equal in size to an electron but having a positive electronic charge. It appears in the structure of a crystal lattice due to the absence of an electron and it will be considered in Chapter 3.

2.2 Bohr theory

The classical theory of Rutherford was based on the fact that the attractive Coulomb force between the positively charged nucleus and the negatively charged electrons was balanced by the outward centrifugal force acting on the revolving electrons. However, in this situation, the electrons being accelerated should radiate energy and the energy loss would cause them to spiral into the nucleus.

In order to resolve this difficulty, Bohr in 1913 proposed the following postulates:

1. The atom has certain energy states for which electrons revolve in stable orbits and do not radiate. The angular momentum of the electron in a stable orbit is an integral multiple of $h/2\pi$, where h is Planck's constant.
2. An electron can move from a low energy state W_1 to a higher energy state W_2 by the absorption of electromagnetic energy, or it can move from a high energy state W_2 to a lower energy state W_1 by radiating electromagnetic energy at a frequency f given by $f = (W_2 - W_1)/h$.

The postulates of Bohr formed the basis of the early quantum theory of matter, which first originated in 1901 when Planck showed that electromagnetic radiation was emitted in quanta according to the equation $W = hf$. Bohr's theory developed this idea further in terms of the quantised energy states and the notion of *quantum numbers*.

Example 1

'The Bohr model of the atom is a concept which aids the visualisation of the basic principles of atomic physics.'

Discuss briefly the limitations of this model as a basis for understanding modern electronic devices.

Assuming that the product of the electron momentum times the circumferential path is an integral multiple of Planck's constant, determine the radii of the first two electron orbits in a hydrogen atom. Calculate also the total energy of an electron in each of these two orbits. Express the energies in electron volts.

(C.E.I.)

Solution

Bohr's model of the atom considered the electron as a particle and emphasised the *orbital* structure of the atom. However, its main limitation is the fact that it did not consider the *wave* property of the electron as developed by Wave Mechanics. Hence, due to the wave-particle duality, emphasis is now placed on energy levels rather than orbits, and this has been found to be very useful in explaining the action of electronic devices such as the semiconductor diode, transistor and maser. Moreover, in some devices like the tunnel diode, 'tunnelling' can be explained quite easily by present day quantum theory.

Problem

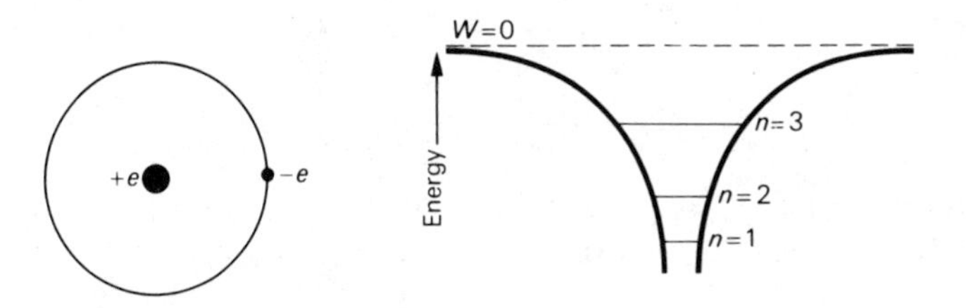

Fig. 6

For the hydrogen atom shown in Fig. 6, the Coulomb force of attraction equals the centrifugal force on the electron. Also, we have $mvr = nh/2\pi$ where $n = 1$ and $n = 2$ for the first and second orbits respectively. Hence

$$\frac{e^2}{4\pi\varepsilon_0 r^2} = \frac{mv^2}{r} = \frac{n^2 h^2}{4\pi^2 m r^3}$$

or

$$r = \frac{n^2 h^2 \varepsilon_0}{\pi m e^2}$$

If r_1, r_2 are the radii of the first and second orbits respectively then

$$r_1 = \frac{h^2 \varepsilon_0}{\pi m e^2}$$

$$r_2 = \frac{4h^2 \varepsilon_0}{\pi m e^2}$$

Now, if W is the total kinetic and potential energy of the atom, the kinetic energy is $\frac{1}{2}mv^2$ for the revolving electron and the potential energy is the work done by the Coulomb force in moving a charge $-e$ from infinity up to a distance r from the nucleus. Hence

$$\text{Kinetic energy} = \tfrac{1}{2}mv^2 = \tfrac{1}{2}\frac{e^2}{4\pi\varepsilon_0 r}$$

$$\text{Potential energy} = \int_{\infty}^{r} \frac{-e^2\,\mathrm{d}r}{4\pi\varepsilon_0 r^2} = -\frac{e^2}{4\pi\varepsilon_0 r}$$

with

$$W = \tfrac{1}{2}\frac{e^2}{4\pi\varepsilon_0 r} - \frac{e^2}{4\pi\varepsilon_0 r} = -\frac{e^2}{8\pi\varepsilon_0 r}$$

If W_1, W_2 are the total energies for the first and second orbits respectively, we obtain

$$W_1 = -\frac{e^2}{8\pi\varepsilon_0 r_1} = -\frac{me^4}{8\varepsilon_0^2 h^2}$$

$$W_2 = -\frac{e^2}{8\pi\varepsilon_0 r_2} = -\frac{me^4}{32\varepsilon_0^2 h^2}$$

Substituting the values $e = 1{\cdot}6 \times 10^{-19}$ C, $m = 9{\cdot}1 \times 10^{-31}$ kg, $h = 6{\cdot}62 \times 10^{-34}$ J-s and $\varepsilon_0 = 8{\cdot}85 \times 10^{-12}\ \mathrm{F\,m^{-1}}$ then yields

$$W_1 = -21{\cdot}62 \times 10^{-19}\ \text{joules} = -13{\cdot}5\ \text{eV}$$
$$W_2 = -\ 5{\cdot}41 \times 10^{-19}\ \text{joules} = -\ 3{\cdot}38\ \text{eV}$$

2.3 Wave Mechanics[7]

Electromagnetic radiation was known to behave both as a wave and as a particle. This wave-particle duality was extended in 1924 by de Broglie to other particles such as the electron, atoms and matter in general. He showed that a particle, such as an electron, with mass m and moving with velocity v can be associated with a wavelength λ given by the de Broglie equation

$$\lambda = h/mv$$

or

$$\lambda = h/p$$

where p is the momentum of the electron and h is Planck's constant.

This result, when applied to atomic electrons, verified Bohr's theory of stable orbits and confirmed that they were given by an integral number of λ such that

$$n\lambda = 2\pi r$$

with

$$\lambda = h/p$$

or

$$nh/p = 2\pi r$$

and

$$pr = nh/2\pi = n\hbar$$

which shows that the angular momentum of the electron is quantised, as postulated by Bohr.

A further development of the wave properties of matter in 1927 was due to Schrödinger and is now known as *Wave Mechanics.* Since the precise behaviour of a particle is unknown, it is expressed as a probability wave in terms of a wave function Ψ which is the solution to Schrödinger's well-known wave equation, derived in Appendix B as

$$\nabla^2\Psi + \frac{8\pi^2 m}{h^2}(W - V)\Psi = 0$$

where W is the total energy, V is the potential energy and ∇^2 is the Laplacian operator.

The uncertainty concerning matter waves was given by Heisenberg's uncertainty principle in 1927. The principle expresses the uncertainty in the position of a particle as Δx and in its momentum as Δp, such that

$$\Delta x \, . \, \Delta p \geqslant \hbar$$

and when expressed as an uncertainty of energy ΔW and of time Δt it becomes

$$\Delta W . \Delta t \geqslant \hbar$$

According to Wave Mechanics, the precise position of an electron cannot be known and so the notion of electron orbits has given way to that of energy levels. The solution to the Schrödinger equation merely expresses the *probability* that an electron occupies a certain energy level, and much of present quantum theory is concerned with the calculation of these energy levels, within the bounds laid down by the uncertainty principle and a further principle due to Pauli in 1925 which is known as the Pauli *exclusion principle.* According to the exclusion principle, no two electrons in an atom can have the same set of quantum numbers and it has led to the determination of the electronic structure of atoms which is described in the next section.

Comment

An alternative approach by Heisenberg using matrices is known as *Matrix*

Mechanics[7] and it yields similar results. Hence, the term *Quantum Mechanics*[8] is sometimes used instead of Wave Mechanics or Matrix Mechanics.

Example 2
A particle with zero potential energy is confined within a potential well of infinite sides and width d as shown in Fig. 7(a). Obtain an expression for its wave functions and show that its energy is quantised.

Solution

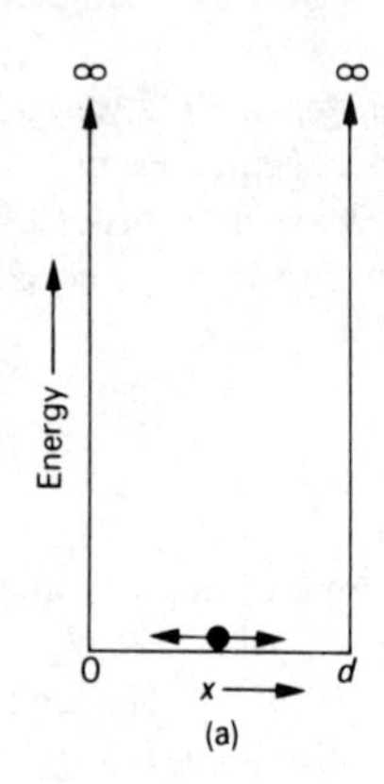

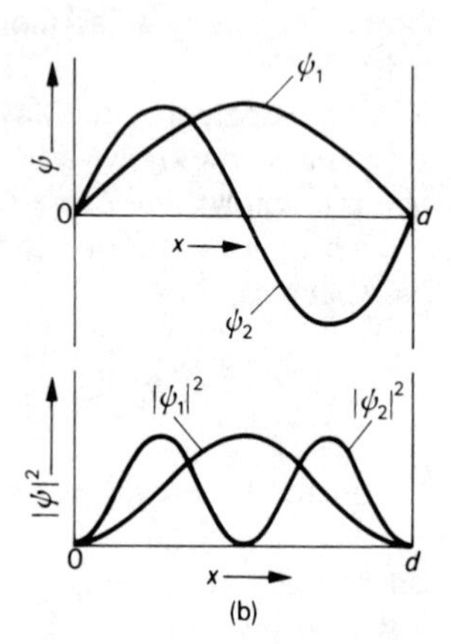

Fig. 7

Assuming distance along the x-axis, the Schrödinger equation for the particle becomes

$$\frac{d^2\Psi}{dx^2} + \frac{8\pi^2 mW\Psi}{h^2} = 0$$

or
$$\frac{d^2\Psi}{dx^2} = -k^2\Psi \qquad \left(k^2 = \frac{8\pi^2 mW}{h^2}\right)$$

This equation represents simple harmonic motion and its solution is of form

$$\Psi = A\sin kx$$

where A is the amplitude of the wave function.

As the particle is *confined* to the well, Ψ must vanish at $x = 0$ and $x = d$. Hence, $\Psi = 0$ when

$$kd = n\pi \qquad (n = 0, 1, 2 \quad \text{etc.})$$

or
$$k = n\pi/d$$

and
$$k^2 = n^2\pi^2/d^2 = \frac{8\pi^2 m W}{h^2}$$

with
$$W = \frac{n^2 h^2}{8md^2}$$

and so the particle energy is quantised as n has integral values only. Furthermore, as the particle must always be somewhere between $x = 0$ and $x = d$ we have

$$\int_0^d \Psi^* \Psi \mathrm{d}x = \int_0^d |\Psi|^2 \mathrm{d}x = 1$$

where $|\Psi|^2 = A^2 \sin^2 kx = \frac{A^2}{2}(1 - \cos 2kx)$. Hence

$$\int_0^d \frac{A^2}{2}(1 - \cos kx)\,\mathrm{d}x = \int_0^d \frac{A^2}{2}\mathrm{d}x = 1$$

or
$$\frac{A^2}{2} d = 1$$

and
$$A = \sqrt{2/d}$$

with
$$\Psi = \sqrt{\frac{2}{d}} \sin \frac{n\pi x}{d}$$

Substituting $n = 1$ and $n = 2$ then yields the respective wave functions Ψ_1 and Ψ_2 shown in Fig. 7(b).

Example 3

An electron beam is incident on a potential barrier of height V_0 and thickness d in the x-direction. Show that electrons can 'tunnel' through the barrier under certain conditions.

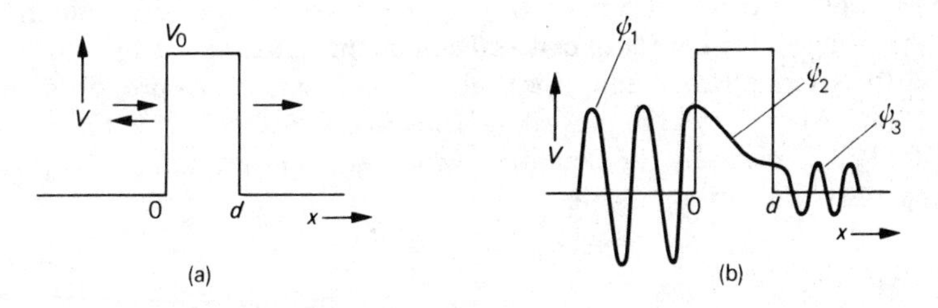

Fig. 8

Solution

The physical model is shown in Fig. 8.

The Schrödinger equation for the incident beam is given by

$$\frac{d^2\Psi}{dx^2}+\frac{8\pi^2 mW}{h^2}\Psi=0$$

or

$$\frac{d^2\Psi}{dx^2}=-k^2\Psi \qquad \left(k^2=\frac{8\pi^2 mW}{h^2}\right)$$

Since a wave can be reflected partially at the barrier, the general solution is of form

$$\Psi_1=Ae^{-jkx}+Be^{jkx}$$

and the total wave function is the sum of the incident and reflected waves.

Inside the barrier for which $0<x<d$, the quantity $(W-V_0)$ becomes negative if V_0 is assumed large. Hence, we obtain

$$\frac{d^2\Psi}{dx^2}=k^2\Psi \qquad \left[k^2=\frac{8\pi^2 m(V_0-W)}{h^2}\right]$$

which has a solution of form

$$\Psi_2=Ce^{-kx}$$

where C is a constant, and so the wave function decreases exponentially through the barrier. At $x=d$, $\Psi=Ce^{-kd}$, and if d is small, Ψ_2 will still be finite. Hence, beyond the barrier for which $V_0=0$, we have again

$$\frac{d^2\Psi}{dx^2}=-k^2\Psi \qquad \left(k^2=\frac{8\pi^2 mW}{h^2}\right)$$

with a solution of form

$$\Psi_3=De^{-jkx}$$

where D is the amplitude of the transmitted wave.

Electrons can therefore 'tunnel' through the barrier instead of surmounting it, depending on the barrier thickness d and on the initial wave amplitude A. If the latter is normalised to unity, the probability of tunnelling is proportional to D^2. Hence, when V_0 is large, tunnelling is possible only if the barrier thickness is small. This is precisely the condition which exists in the tunnel diode and tunnelling is therefore possible.

2.4 The Periodic Table[6]

The first attempt to group elements together on the basis of atomic weights was made in 1869 by Mendeleef. He arranged elements according to their chemical similarity and increase in atomic weight. This took the form of a *periodic table* consisting of eight *Groups*.

The arrangement of groups led to the discovery of many new elements which the table predicted, and much of its form is still retained. However, there were some difficulties which could not be resolved by the table and it was later developed by Bohr into its present form.

The periodic table of Bohr was arranged on the basis of increasing *atomic number* and so was built around an electronic structure. According to this structure, electrons are arranged in *shells* which consist of one or more *sub-shells.* Their arrangement is determined by various quantum numbers such as the principal quantum number n, the orbital quantum number l, the magnetic quantum number m and the spin quantum number s.

The shells are called the K, L, M, N . . . shells for which $n = 1, 2, 3$ etc. Sub-shells are designated s, p, d, f . . . for which $l = 0, 1, 2$ etc. According to Pauli's exclusion principle no two electrons can have the same set of quantum numbers. However, with opposite spins the spin quantum number can be $\pm\frac{1}{2}$. Electrons in shells for which $n = 1$ and $l = 0, 1, 2$ etc. are designated 1s, 1p, 1d . . ., while those in shells for which $n = 2$ and $l = 0, 1, 2$ etc. are designated 2s, 2p, 2d . . . and so on for other values of n. A typical arrangement of the electron structure is given in Table 1 and exemplified for the three elements He, C and Cu in Fig. 9. This simple picture, however, breaks down for the heavier atoms.[6]

Table 1

shell designation	*quantum numbers* n	l	m	s	*sub-shell designation*	*maximum number of electrons*
K	1	0	0	$\pm\frac{1}{2}$	1s 1s	1 1
L	2	0 1	0, ±1	$\pm\frac{1}{2}$	2s 2p	2 6
M	3	0 1 2	0, ±1, ±2	$\pm\frac{1}{2}$	3s 3p 3d	2 6 10
N	4	0 1 2 3	0, ±1 ±2, ±3	$\pm\frac{1}{2}$	4s 4p 4d 4f	2 6 10 14

The elements, when grouped according to the Bohr table, show reasonable similarity with Mendeleef's table. The same eight groups can be observed and the previous difficulties have been resolved. Generally, the elements fall into the following main categories: (i) the metals (ii) the non-metals (iii) the semi-metals or semiconductors. Some details of the periodic table appear in Table 2 and a list of the elements is given in Appendix A.

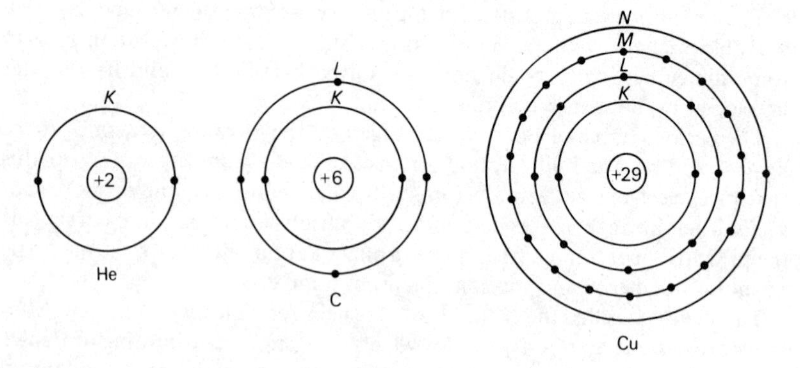

Fig. 9

Table 2

	Group							
shell	*1*	*2*	*3*	*4*	*5*	*6*	*7*	*8*
K	H							
L	Li	Be	B	C		O		
M	Na	Mg	Al	Si	P			
N	Cu	Zn	Ga	Ge	As	Se	Mn	Fe, Co, Ni
O	Ag	Cd	In	Sn	Sb	Te	I	
P	Cs	Ba			Bi	W		
Q				Th				

3

Semiconductor theory

Solid materials like copper, silver or gold conduct electricity easily and are called good *conductors*. They have a conductivity at room temperature around $10^7\,S\,m^{-1}$. In contrast, solid materials such as glass, mica or porcelain are extremely poor conductors of electricity and are called *insulators*. Their conductivity is very low and around $10^{-16}\,S\,m^{-1}$. However, there are also some materials such as germanium and silicon whose conductivity lies in between those of conductors and insulators and are called *semiconductors*. They form an extremely important group of materials which have been studied in great detail over recent years.

3.1 Conductors and insulators

The atoms of many solid materials are arranged in the form of a regular structure called a crystal lattice.[9] A common example in metals is the cubic lattice and there are two forms which are known as the *body centred cube* and the *face centred cube*. The atoms are positioned at certain points in the crystal lattice as shown in Fig. 10.

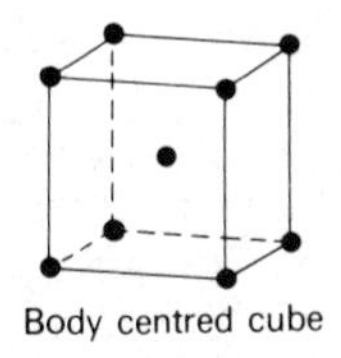

Body centred cube

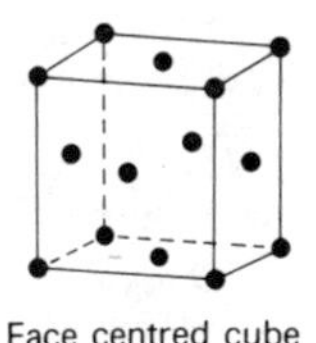

Face centred cube

Fig. 10

In a single isolated atom the energy levels are clearly separated as shown in Fig. 11(a), where the horizontal lines represent the quantised energy levels of the electron orbits, and their diameters are given by the length of the lines. When several atoms come together to form a crystal, there is coupling between the atoms, and as the atomic spacing decreases, the energy levels split and overlap as shown in Fig. 11(b). Hence, many more energy levels are possible and they become very closely spaced to form *energy bands* as shown in Fig. 11(c) for the simple case of two atoms.

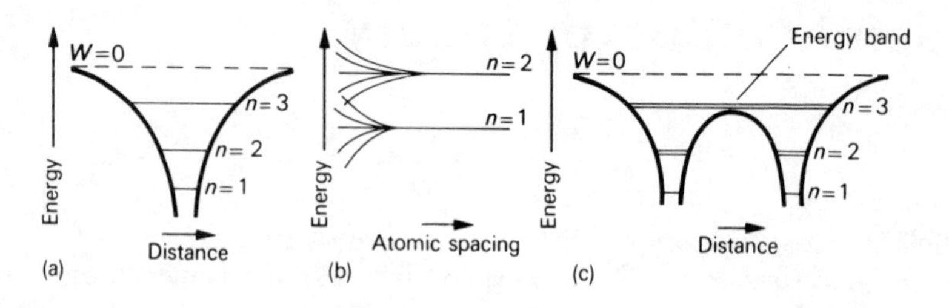

Fig. 11

At 0 K the electrons are in the lowest energy levels which are all filled. This is the *valence band*. For conduction to take place, electrons must be raised into unfilled energy levels which exist in the *conduction band*. It requires an applied electric field which can raise the electrons above the forbidden *energy gap*, into the conduction band. Such valence electrons are then free to move about from atom to atom and give rise to a current flow.

In an insulator, the conduction band and valence band are separated by a large forbidden energy gap. The gap is about 7 eV for an insulator like diamond, and a few eV for other insulators. At normal temperatures and applied fields, the probability of valence electrons acquiring sufficient energy to overcome the energy gap is negligible. Hence, a current cannot flow readily because virtually no electrons are present in the conduction band. This is illustrated in Fig. 12.

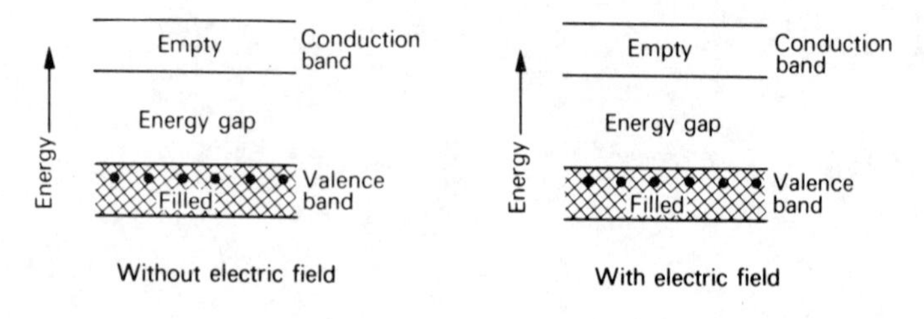

Fig. 12

In a conductor, the conduction band and valence band are very close to one another or may even overlap. Due to thermal vibration, the conduction band is partially filled with valence electrons which move about randomly. With an applied electric field, electrons at the top of the valence band acquire sufficient energy to raise them into the conduction band, and they also move with a drift velocity in one direction. These valence electrons are readily available for conduction as a flow of current through the conductor. This is illustrated in Fig. 13.

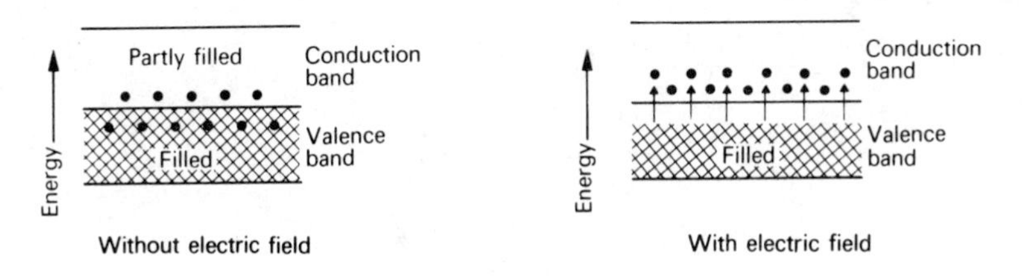

Fig. 13

3.2 Semiconductors

In some materials like germanium and silicon, their small energy gap of about 1 eV at room temperature makes them partial conductors. Hence, they are called *semiconductors* and there are two types known respectively as *intrinsic* semiconductors and *extrinsic* semiconductors.

Intrinsic semiconductors

The electronic structure of germanium and silicon atoms is shown in Fig. 14 and both atoms are tetravalent because they have four valence electrons in the outermost orbit. The valence electrons of neighbouring atoms form *covalent* bonds in which the valence electrons are tightly bound. Hence, at 0 K pure germanium and pure silicon behave as insulators.

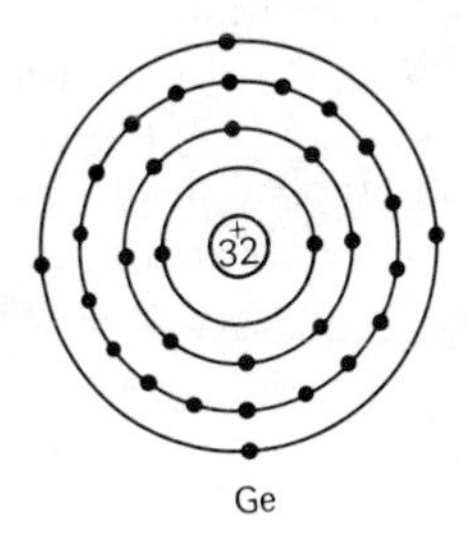

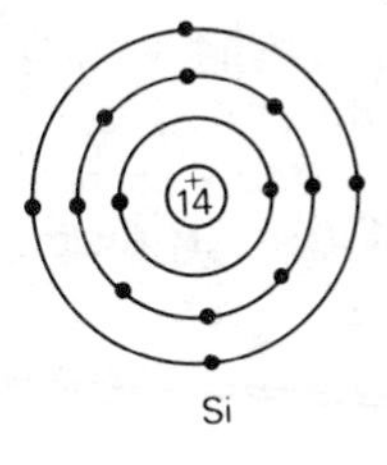

Fig. 14

At room temperature however, some of the bonds are broken by thermal agitation and a few electrons move into the conduction band. The breaking of a covalent bond moves an electron into the conduction band, thus creating a *hole* or vacancy in the valence band. Hence, electron-hole pairs are produced throughout the material, whose conductivity is then slightly increased. The material is called an *intrinsic* semiconductor, and the process is illustrated in Fig. 15 for silicon.

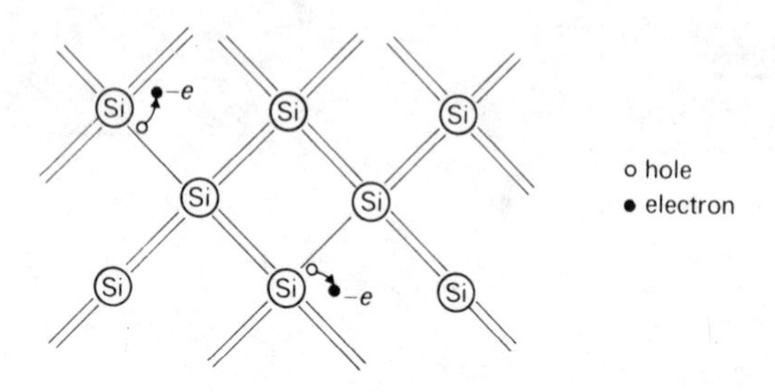

Fig. 15

Due to thermal agitation, new electron-hole pairs are continuously produced while others recombine and are annihilated, thus maintaining constant charge density at a given temperature. Hence, the electron carrier density n and the hole carrier density p are equal such that

$$n = p = n_i$$

and

$$n \cdot p = n_i^2$$

where n_i is the carrier density of intrinsic material. As n_i is a function of temperature, the maximum operating temperature for germanium should be $< 100°C$ and for silicon $< 200°C$.

Electrons and holes with thermal energies move about randomly through the lattice structure. Since hole movement is due to an electron moving into a hole position, under an applied field E, electrons and holes acquire an additional *drift* velocity and move in opposite directions along the applied field E.

If the drift velocities of electrons and holes are $-v_n$ and v_p respectively, the *mobility factor* is defined as the velocity per unit applied field (m s^{-1}/V-m). Hence, if μ_n, μ_p are the mobility factors respectively, we have $\mu_n = -v_n/E$ and $\mu_p = v_p/E$

or

$$|v_n| = \mu_n E$$

$$|v_p| = \mu_p E$$

However, the mobility of holes is less than that of electrons, because a hole movement in one direction is due to several electrons moving in the opposite direction, as each electron has to move in turn into the hole position. Nevertheless both holes and electrons contribute to the conduction process. As n and p are the densities of electrons and holes respectively the total charge flow q due to an applied field E becomes

$$q = ne|v_n| + pe|v_p| = ne\mu_n E + pe\mu_p E$$

or

$$q = E(ne\mu_n + pe\mu_p)$$

where e is the electronic or hole charge.

If the conductivity of the material is σ, it is given by the charge flow per unit applied field E and so we have

$$\sigma = q/E$$

or

$$\sigma = (ne\mu_n + pe\mu_p) \qquad \mathrm{S\,m^{-1}}$$

Extrinsic semiconductors

A semiconductor such as pure germanium or pure silicon has an impurity of about one part in 10^{10}. However, if an impurity is added even as a small amount of one part in 10^7, the conductivity of the material can change considerably by about a factor of 10^3. This is because the impurity atoms enter into the lattice structure and produce many extra charge carriers. The process is called *doping* and the doped material is called an *extrinsic* semiconductor.

In the case of tetravalent silicon, addition of trivalent indium produces one unfilled bond which creates a *hole* and the semiconductor is called *p-type* material, while the addition of pentavalent antimony produces an extra *electron* and the semiconductor is called *n-type* material. Substances like indium are called *acceptors* while those like antimony are called *donors*. This is shown in Fig. 16.

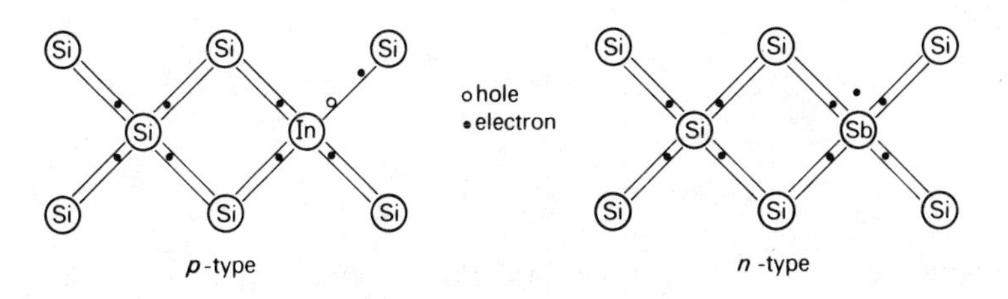

Fig. 16

In extrinsic semiconductors, the carrier concentrations are dependent on the impurity added. Donors such as As, Sb and P produce a donor level near the *conduction* band and due to thermal excitation, electrons move easily from this level to the conduction band thereby increasing the electrons in the conduction band. Acceptors such as Ga, Al and In produce an acceptor level near the *valence* band and electrons move easily from the valence band to this level, thereby increasing the holes in the valence band. This is illustrated in Fig. 17.

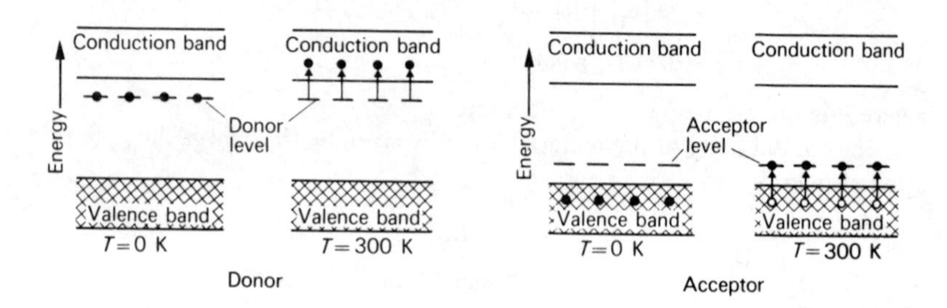

Fig. 17

The carrier densities in the extrinsic material will depend on the concentration of the impurity atoms introduced by doping, and since the material must remain electrically neutral after doping, the number of positive and negative charges must be equal. If N_a, N_d are the concentrations of impurity atoms in acceptor and donor materials, which are assumed to be fully ionised, then

$$n = p + N_d \qquad (n\text{-type material})$$

$$p = n + N_a \qquad (p\text{-type material})$$

Since doping hardly affects the production and recombination of electron-hole pairs we also have $n \, . \, p = n_i^2$. Hence, for n-type material we obtain

$$n_i^2/p = p + N_d$$

or

$$p^2 + pN_d - n_i^2 = 0$$

with

$$p = \frac{-N_d + \sqrt{N_d^2 + 4n_i^2}}{2}$$

or

$$p = \frac{-N_d}{2} + \frac{N_d}{2}\left[1 + \left(\frac{2n_i}{N_d}\right)^2\right]^{1/2}$$

Since $N_d \gg n_i$ in n-type material, the bionomial approximation yields

$$p \simeq -N_d/2 + N_d/2 + n_i^2/N_d$$

or

$$p \simeq n_i^2/N_d$$

and
$$n = p + N_d = \frac{n_i^2 + N_d^2}{N_d}$$

or
$$n \simeq N_d$$

For p-type material the analysis is similar and we obtain

$$n \simeq n_i^2/N_a$$

and
$$p \simeq N_a$$

The charge carriers in p-type material are mainly holes and are called the *majority carriers*, while the electrons are the *minority carriers*. In n-type material, electrons are the *majority carriers* and holes are the *minority carriers*.

Example 4

Explain the process of electrical conduction in (a) an intrinsic semiconductor at room temperature (b) a heavily doped semiconductor, indicating briefly how the conductivity depends on temperature in each case.

An intrinsic silicon specimen at approximately 300 K has a conductivity of $4{\cdot}3 \times 10^{-4}$ ohm^{-1} m^{-1}. What is the intrinsic carrier concentration? If a current is passed through the specimen what proportion of it is carried by the electrons? The same specimen is now doped to make n-type. The donor concentration is 10^{21}/m^3. Find the hole density of the doped specimen and also the proportion of current that would now be carried by electrons. Assume that the mobilities are substantially unchanged by the doping process.

Mobility of electrons in silicon at 300 K = $0{\cdot}135$ m^2/V-s
Mobility of holes in silicon at 300 K = $0{\cdot}048$ m^2/V-s (U.L.)

Solution

The conduction process in intrinsic and extrinsic semiconductors was described in Section 3.2. Furthermore, it can be shown (see Appendix C) that the relationship between carrier density and temperature is given by

$$n = p = AT^{3/2}e^{-\Delta V/2kT}$$

where A is a constant and ΔV is the width of the forbidden energy gap in eV.

A plot of n for intrinsic material is shown in Fig. 18 and it will be observed that at low temperatures, n is small, i.e. low conductivity, while at room temperature ($T = 300$ K), the conductivity is much larger, especially in doped material. At high temperatures, however, n increases considerably and can cause a further increase in temperature due to current flow, with possible damage to the material.

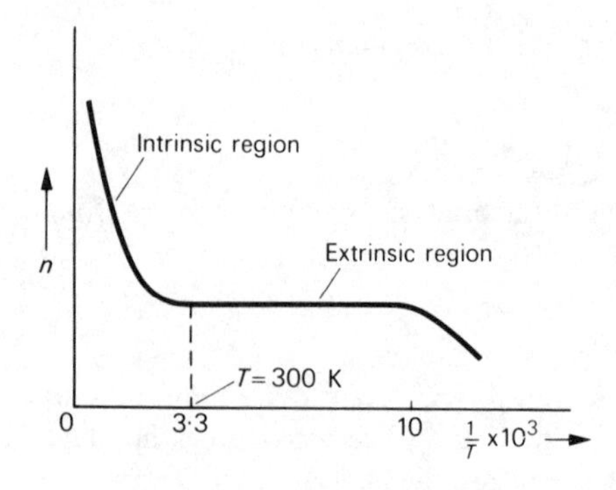

Fig. 18

Problem

For intrinsic material we have $\sigma = e(n\mu_n + p\mu_p)$, with $n = p$ at 300 K. Hence, we obtain

$$n = \frac{\sigma}{e(\mu_n + \mu_p)} = \frac{4{\cdot}3 \times 10^{-4}}{1{\cdot}6 \times 10^{-19}(0{\cdot}135 + 0{\cdot}048)}$$

or

$$n = 1{\cdot}47 \times 10^{16}/\text{m}^3$$

The proportion of current carried by electrons is given by

$$\frac{n\mu_n}{(n\mu_n + p\mu_p)} = \frac{\mu_n}{(\mu_n + \mu_p)} = \frac{0{\cdot}135}{0{\cdot}183} = 0{\cdot}738$$

For extrinsic n-type material we have

$$p = n_i^2/N_d = \frac{(1{\cdot}47 \times 10^{16})^2}{10^{21}} = 2{\cdot}16 \times 10^{11}/\text{m}^3$$

with

$$n \simeq N_d \simeq 10^{21}/\text{m}^3$$

Hence, the proportion of current carried by the electrons is given by

$$\frac{n\mu_n}{(n\mu_n + p\mu_p)} \simeq 1 - p\mu_p/n\mu_n \simeq 1 - \frac{2{\cdot}16 \times 10^{11} \times 0{\cdot}048}{10^{21} \times 0{\cdot}135}$$

which is $\simeq (1 - 10^{-10})$ and so the current is mainly due to electrons.

3.3 Fermi-Dirac distribution[10]

The statistical distribution of electrons in a solid can be determined by Fermi-Dirac statistics which is based upon quantum principles. The probability $P(W)$ of an electron having an energy between W and $(W + \mathrm{d}W)$ is given by the

Fermi-Dirac distribution function

$$P(W) = \frac{1}{1 + e^{(W - W_F)/kT}}$$

where k is Boltzmann's constant, T is the absolute temperature and W_F is the well-known *Fermi level* of the material which gives a probability of 0·5 that an electron has an energy $W = W_F$. This is illustrated in Fig. 19.

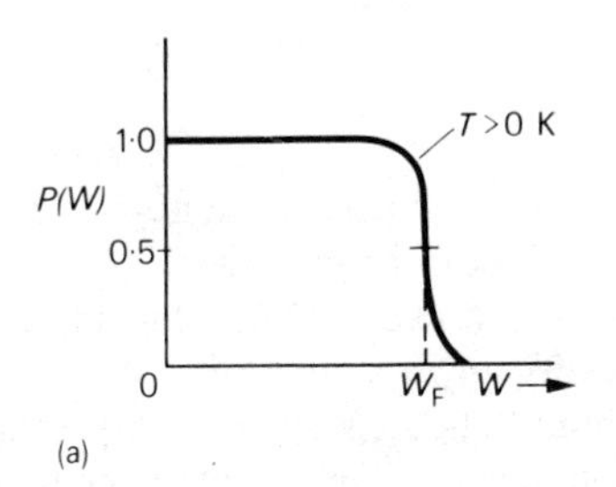

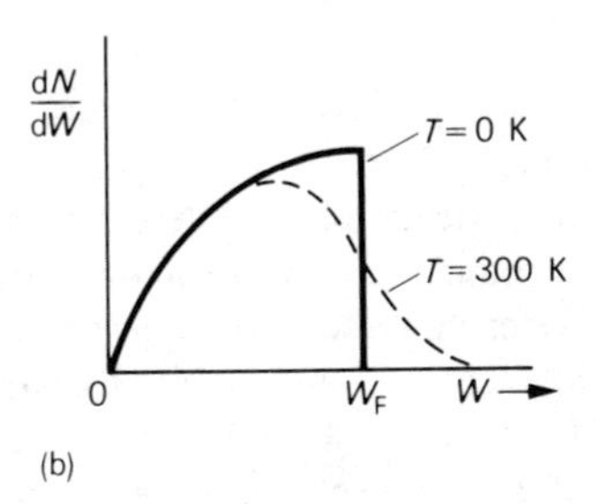

Fig. 19

From Fig. 19 it will be observed that $P(W) = 1$ when $W \ll W_F$, i.e. all the lower energy levels are filled, and $P(W) = 0$ when $W \gg W_F$, i.e. all the higher energy levels are empty. Hence, the density of electrons and holes in the material can be calculated at various temperatures. In Appendix C it is shown that the density of electrons in the conduction band at temperature T is given by

$$n = 2\left(\frac{2\pi m_e^* kT}{h^2}\right)^{3/2} e^{-(W_c - W_F)/kT}$$

where m_e^* is the effective electron mass, h is Planck's constant and W_c is the lowest energy level in the conduction band. Similarly, the density of holes is given by

$$p = 2\left(\frac{2\pi m_h^* kT}{h^2}\right)^{3/2} e^{-(W_F - W_v)/kT}$$

where m_h^* is the effective hole mass and W_v is the highest energy level in the valence band.

In an intrinsic semiconductor, the density of electrons equals the density of holes and the Fermi level which has a probability of 0·5 must lie mid-way between the valence band and the conduction band. In n-type extrinsic material which has a large density of electrons, W_F lies above the donor level, while in p-type material with a large density of holes, W_F lies below the acceptor level as shown in Fig. 20. However, with low doping or at high temperatures, W_F tends to approach its mid-gap position.

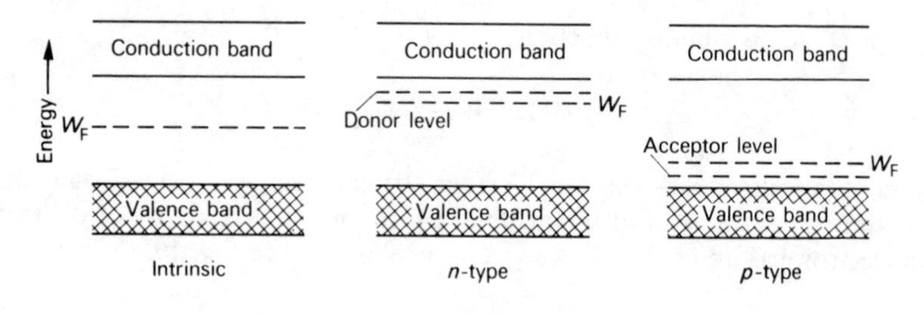

Fig. 20

Example 5

Give a brief qualitative description of the means by which current flows in an *n*-type semiconductor, explaining the origins of the charge carriers.

Determine the Fermi energy in *n*-type germanium at 300 K, if the impurity content is one part in 10^6. Explain and justify any assumptions.

There are $4{\cdot}4 \times 10^{28}$ atoms/m^3 in germanium. The constant in the expression relating the number of electrons per unit volume in the conduction band to the temperature and energy levels is $4{\cdot}83 \times 10^{21}$ m^{-3} (K)$^{-3/2}$. The difference in energy between the bottom of the conduction band and the top of the valence band is 0·72 eV, between the bottom of the conduction band and the donor level, 0·01 eV. (U.L.)

Solution

The answer to the first part will be found in Section 3.2 under the heading 'Extrinsic semiconductors'.

Problem

The assumptions made are as follows:

1. All the donor atoms are ionised at 300 K so the probability of occupancy of the donor level is zero.
2. Since the donor level is only 0·01 eV below the conduction band and has an occupancy assumed to be zero, the Fermi level must lie below the donor level.

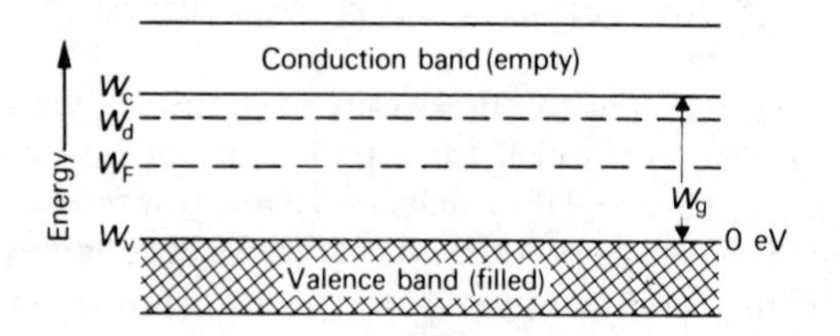

Fig. 21

The energy level diagram is shown in Fig. 21 with W_v assumed arbitrarily as 0 eV and W_g as the energy gap between the top of the valence band and the bottom of the conduction band. From Section 3.3 we have

$$n = 2\left(\frac{2\pi m_e^* kT}{h^2}\right)^{3/2} \mathrm{e}^{-(W_c - W_F)/kT}$$

or

$$n = AT^{3/2}\,\mathrm{e}^{-(W_c - W_F)/kT}$$

where A is a constant of proportionality. Also, we have

$$n \simeq N_d \simeq \frac{4{\cdot}4 \times 10^{28}}{10^6} \simeq 4{\cdot}4 \times 10^{22}\ \text{electrons/m}^3$$

with

$$A = 4{\cdot}83 \times 10^{21}\ \mathrm{m}^{-3}\,(\mathrm{K})^{-3/2}$$

Hence, we obtain

$$4{\cdot}4 \times 10^{22} = 4{\cdot}83 \times 10^{21} \times 300^{3/2}\,\mathrm{e}^{-(W_g - W_F)/kT}$$

or

$$(W_F - W_g) = kT \ln\left[\frac{4{\cdot}4 \times 10^{22}}{4{\cdot}83 \times 10^{21} \times 5{\cdot}2 \times 10^3}\right] \text{joules}$$

and converting joules to electron volts yields

$$W_F = 0{\cdot}72 + \frac{1{\cdot}38 \times 10^{-23} \times 300}{1{\cdot}6 \times 10^{-19}} \times 2{\cdot}3 \times \log_{10}(1/570)$$

or

$$W_F = 0{\cdot}72 - 0{\cdot}16 = 0{\cdot}56\,\text{eV}$$

The Fermi level is therefore 0·56 eV above the highest level of the valence band or 0·16 eV below the lowest level of the conduction band.

3.4 Hall effect[10]

If a voltage is applied to the slab of p-type material shown in Fig. 22 holes will drift in the x-direction. When a perpendicular magnetic field **B** is applied to the slab in the z-direction, the force acting on the holes will cause them to drift to one side of the slab. It establishes an electric field E across the slab and a voltage $V_H = -E_y d$, where d is the width of the slab. This is the *Hall effect* and V_H is the Hall voltage acting in the *negative* y-direction.

The force acting on the holes is given by

$$\mathbf{F} = e(\mathbf{E} + \mathbf{v} \times \mathbf{B})$$

where e is the hole charge, **E** is the total electric field acting on the holes and **v** is the drift velocity of the holes along the slab. Equilibrium is obtained when we have $\mathbf{F} = \mathbf{0}$. Hence

$$\mathbf{E} = -(\mathbf{v} \times \mathbf{B}) = \mathbf{B} \times \mathbf{v}$$

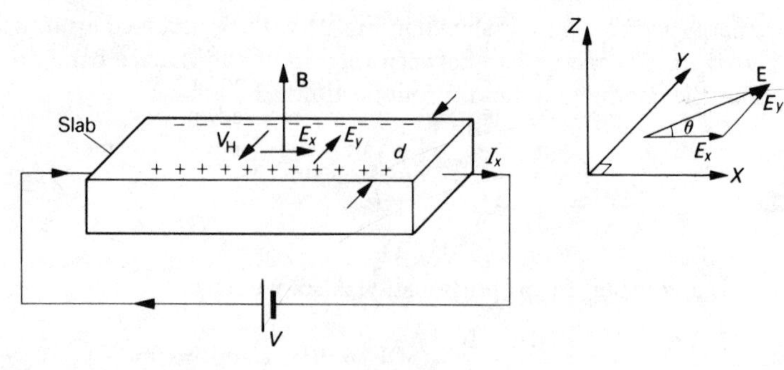

Fig. 22

and is a vector acting in the y-direction. Here

$$\mathbf{E} = \mathbf{E}_y = B_z v_x \mathbf{y}$$

where $\mathbf{y}$ is the unit vector in the positive y-direction.

If the current density is J due to a hole density p then $J_x = pev_x$ and we obtain

$$E_y = \frac{J_x}{pe} B_z = R_H J_x B_z$$

where $R_H = 1/pe$ and is known as the *Hall coefficient*.

Since the mobility of holes is given by

$$\mu_p = v_x/E_x = \frac{pev_x}{peE_x}$$

or

$$\mu_p = R_H J_x/E_x = R_H/\rho \qquad (J_x = pev_x)$$

where $\rho = E_x/J_x$ is the resistivity of the material, the mobility of holes may be determined by making measurements of ρ and R_H. Alternatively, the mobility of electrons may be determined if an n-type slab is used.

The total electric field across the slab is the vector sum of E_x and E_y, and acts at the Hall angle θ which is given by

$$\tan\theta = \frac{E_y}{E_x} = \frac{R_H J_x B_x}{J_x \rho} = \frac{R_H B_z}{\rho}$$

or

$$\tan\theta = \mu_p B_z$$

Example 6

The sample of semiconductor material illustrated in Fig. 23 has dimensions 5 mm, 2 mm and 4 mm in the x-, y- and z-directions respectively. A 1·5 V supply results in a current $I = 35$ mA. With $B_y = 0{\cdot}09$ T, the high-impedance voltmeter V indicates 14 mV with the polarity shown.

Determine the sign of the carriers, their mobility and density of doping.

Derive all formulae used and mention all assumptions.

What would be the effect of using a bar of the same dimensions but doped with the same number of carriers of the opposite polarity?

Discuss briefly how this phenomenon may be used in practice either (a) in magnetic field measurements or (b) in audio-frequency power measurements. (C.E.I.)

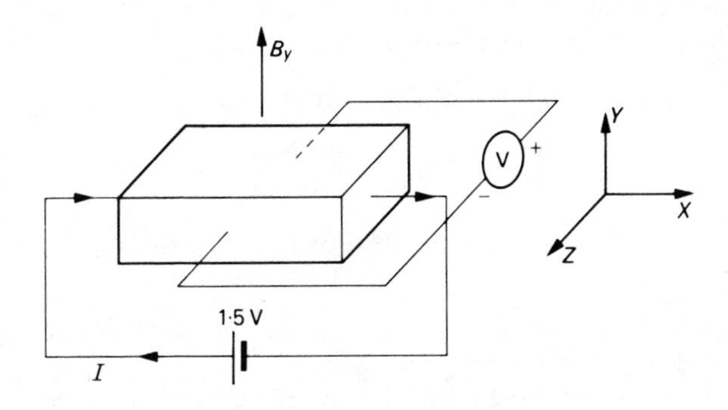

Fig. 23

Solution

Since the Hall voltage acts in the *negative* z-direction, the charge carriers are electrons and the sample is of the n-type.

The relevant equations were derived in Section 3.4, and for the geometry shown in Fig. 23 we have

$$E_z = R_H j_x B_y$$

$$R_H = 1/ne$$

$$\mu_n = R_H/\rho$$

where n is the electron doping density, μ_n is the electron mobility and ρ is the resistivity of the sample. Hence

$$E_z = \frac{14 \times 10^{-3}}{4 \times 10^{-3}} = 3{\cdot}5\ \mathrm{V\,m^{-1}}$$

$$J_x = \frac{35 \times 10^{-3}}{4 \times 2 \times 10^{-6}} = 4{\cdot}375 \times 10^3\ \mathrm{A\,m^{-2}}$$

Also

$$R_H = E_z/J_x B_y = \frac{3{\cdot}5}{4{\cdot}375 \times 10^3 \times 9 \times 10^{-2}}$$

or

$$R_H = 8{\cdot}9 \times 10^{-3}$$

Hence $$n = 1/R_{\mathrm{H}}e = \frac{1}{8{\cdot}9 \times 10^{-3} \times 1{\cdot}602 \times 10^{-19}}$$

or $$n = 7{\cdot}01 \times 10^{20} \text{ electrons/m}^3$$

Also $$\rho = \frac{V}{I}\frac{A}{l} = \frac{1{\cdot}5 \times 2 \times 4 \times 10^{-6}}{35 \times 10^{-3} \times 5 \times 10^{-3}}$$

where V is the applied voltage, I is the input current, l is the length and A is the cross-sectional area of the sample. Hence

$$\rho = 6{\cdot}86 \times 10^{-2} \text{ ohm-m}$$

with $$\mu_n = R_{\mathrm{H}}/\rho = \frac{8{\cdot}9 \times 10^{-3}}{6{\cdot}86 \times 10^{-2}}$$

or $$\mu_n = 0{\cdot}130 \text{ m}^2/\text{V-s}$$

Assumptions
1. The donor atoms are completely ionised.
2. There is no temperature gradient across the sample due to the current flow and the temperature is assumed to be 300 K.
3. The sample contains mainly one type of charge carriers, i.e. the doping level is high.

Comment
If the bar were doped with carriers of the opposite polarity, the Hall voltage would be reversed in direction but unchanged in magnitude.

Applications
For magnetic field measurements, the sample material is mounted at the end of a 'probe' which is placed in the centre of the magnetic field to be measured. The Hall voltage generated is amplified and recorded on a meter. The latter may be calibrated directly in teslas.

For audio power measurements, the external d.c. voltage is replaced by an audio voltage source. The alternating Hall voltage produced is amplified and recorded on a meter. The latter may be calibrated directly in watts.

3.5 p-n junction

A *p-n* junction is formed when a *p*-type semiconductor is in contact with an *n*-type semiconductor as shown in Fig. 24. The junction is either of the grown, alloyed or diffused type, whereby the crystal structure is maintained throughout the junction region and is illustrated in Fig. 24.

When a *p-n* junction is formed, holes in the *p*-type material diffuse across the junction into the *n*-type material. Likewise, electrons in the *n*-type material

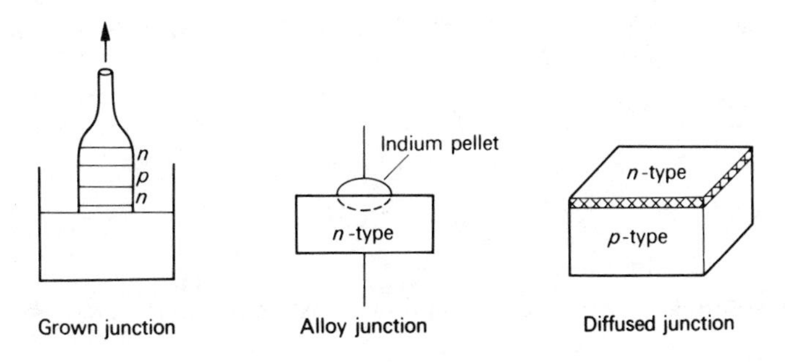

Fig. 24

diffuse into the p-type material. Due to the fixed charged ions on either side of the junction, the p-type material acquires a negative potential and the n-type material acquires a positive potential, thus creating a region of space-charge at the junction. The diffusion process therefore produces a potential barrier or *contact potential* ϕ at the junction, as shown in Fig. 25. This potential barrier prevents any further diffusion of charge carriers across the junction and so a depletion region which is depleted of charge carriers is formed at the junction.

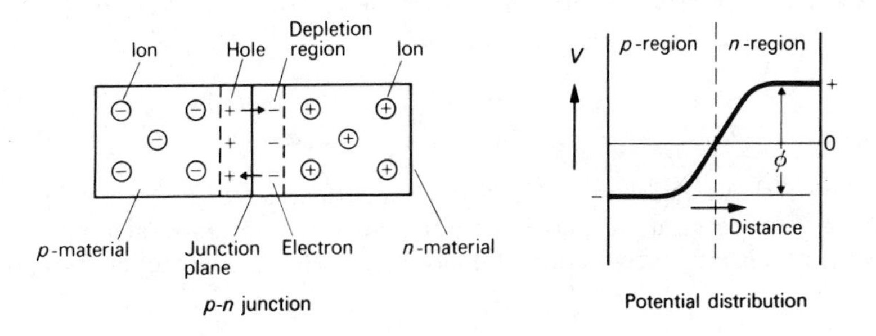

Fig. 25

The contact potential ϕ produces an electric field across the junction which causes some minority holes to drift into the p-type region and some minority electrons to drift into the n-type region. This *drift* current must equal the *diffusion* current and in equilibrium, the net current across the junction is zero. For the case of holes moving across the junction, the drift current density due to an electric field E across the junction is $\mu_p peE$ where p is the density of holes with charge e and mobility μ_p. The diffusion current density is due to holes diffusing from a region of high hole density to one of low hole density, i.e. p decreases with x increasing. It is therefore proportional to the *rate* of change or slope $\mathrm{d}p/\mathrm{d}x$ which is negative. The constant of proportionality is defined as the diffusion coefficient D_p and the diffusion current density for holes is $-D_p \mathrm{d}p/\mathrm{d}x$. Hence, we have

$$e[\mu_p p E - D_p \mathrm{d}p/\mathrm{d}x] = 0$$

or

$$\mu_p p E = D_p \mathrm{d}p/\mathrm{d}x$$

and

$$\frac{\mu_p E}{D_p} = \frac{1}{p}\frac{\mathrm{d}p}{\mathrm{d}x}$$

Substituting for $E = -\mathrm{d}V/\mathrm{d}x$ and the Einstein relation* $\mu_p/D_p = e/kT$ we obtain

$$\frac{e}{kT}\frac{\mathrm{d}V}{\mathrm{d}x} = -\frac{1}{p}\frac{\mathrm{d}p}{\mathrm{d}x}$$

If V_p and V_n are the potentials on either side of the junctions and p_p and p_n are the corresponding hole densities, integration then yields

$$(V_n - V_p) = \phi = \frac{kT}{e}\ln\left(\frac{p_p}{p_n}\right)$$

or

$$p_p/p_n = \mathrm{e}^{e\phi/kT}$$

where ϕ is the contact potential.

At equilibrium we also have $p_p n_p = n_\mathrm{i}^2 = p_n n_n$ and so the electron densities on either side of the junction are given by

$$n_n/n_p = \mathrm{e}^{e\phi/kT}$$

or

$$p_p/p_n = n_n/n_p = \mathrm{e}^{e\phi/kT}$$

It was shown in Section 3.2 that for *p*-type material we have $p_p \simeq N_\mathrm{a}$ where N_a is the concentration of acceptor atoms and for *n*-type material $p_n \simeq n_\mathrm{i}^2/N_\mathrm{d}$ where N_d is the concentration of donor atoms. Hence, substituting for p_p/p_n yields

$$\frac{p_p}{p_n} = \mathrm{e}^{e\phi/kT} \simeq \frac{N_\mathrm{a} N_\mathrm{d}}{n_\mathrm{i}^2}$$

or

$$e\phi/kT \simeq \ln(N_\mathrm{a} N_\mathrm{d}/n_\mathrm{i}^2)$$

and

$$\phi \simeq \frac{kT}{e}\ln\left(\frac{N_\mathrm{a} N_\mathrm{d}}{n_\mathrm{i}^2}\right)$$

which is the height of the barrier or contact potential at the junction during equilibrium conditions.

* See Example 7

Example 7

An expression for the total electron current density in a semiconductor is

$$\left[e\mu nE + eD\frac{\mathrm{d}n}{\mathrm{d}x}\right]$$

Describe the physical processes that account for each term and discuss the equilibrium condition.

Hence or otherwise derive the Einstein relationship between the diffusivity and the mobility of the electrons.

A specimen of *p*-type indium antimonide has an electron mobility of $6{\cdot}2\,\mathrm{m^2/V\text{-}s}$ at 290 K. If the lifetime of the minority charge carriers is 3×10^{-8} s, calculate their diffusion length. (U.L.)

Solution

The answer to the first part of the question was given at the beginning of this section. For electrons, use μ_n, n and $D_n\mathrm{d}n/\mathrm{d}x$ instead of μ_p, p and $-D_p\mathrm{d}p/\mathrm{d}x$ respectively.

Einstein relationship

In equilibrium, the total electron current density must be zero. Hence

$$0 = e\mu nE + eD\frac{\mathrm{d}n}{\mathrm{d}x}$$

or

$$\frac{\mathrm{d}n}{\mathrm{d}x} = \frac{-n\mu E}{D}$$

Previously it was shown that the density of minority electrons is given by

$$n_n = n_p \mathrm{e}^{e\phi/kT}$$

or

$$n = A\mathrm{e}^{qV/kT}$$

if $n = n_n$, $e\phi = qV$ and $n_p = A$ which is constant at a given temperature. Hence differentiating with respect to x yields

$$\frac{\mathrm{d}n}{\mathrm{d}x} = A\mathrm{e}^{qV/kT}\frac{q}{kT}\frac{\mathrm{d}V}{\mathrm{d}x}$$

or

$$\frac{\mathrm{d}n}{\mathrm{d}x} = -\frac{nq}{kT}E \qquad \left(E = -\frac{\mathrm{d}V}{\mathrm{d}x}\right)$$

Substituting for $\mathrm{d}n/\mathrm{d}x$ in the equation above we obtain

$$-\frac{nq}{kT}E = -\frac{n\mu E}{D}$$

or

$$\mu/D = q/kT$$

which is the Einstein relationship required.

Furthermore, if L is the diffusion length with no applied electric field then[11]

$$L = \sqrt{D\tau}$$

or

$$L = \sqrt{\mu kT\tau/q}$$

where τ is the carrier lifetime. Hence

$$L = \sqrt{\frac{6{\cdot}2 \times 1{\cdot}38 \times 10^{-23} \times 290 \times 3 \times 10^{-8}}{1{\cdot}6 \times 10^{-19}}}$$

or

$$L = 6{\cdot}8 \times 10^{-5}\ \text{m}$$

Example 8

A *p-n* junction is made of intrinsic germanium, with 10^{13} free electrons per cm^3 doped with 10^{17} and 5×10^{16} ionised impurity atoms per cm^3 on the *p*- and *n*-sides respectively. The diffusion constants for the minority electrons and holes are 100 and 50 $\text{cm}^2\,\text{s}^{-1}$ respectively, the diffusion lengths being 0·08 cm in each case.

Estimate the value of the energy barrier assuming that the majority concentration in the material is proportional to $e^{-W/kT}$ where W is the energy difference between the Fermi level and the conduction band (for electrons) or the valence band (for holes).

Calculate the saturation current density assuming that the deviation from the equilibrium concentration of minority carriers varies exponentially with distance from the junction. (U.L.)

Solution

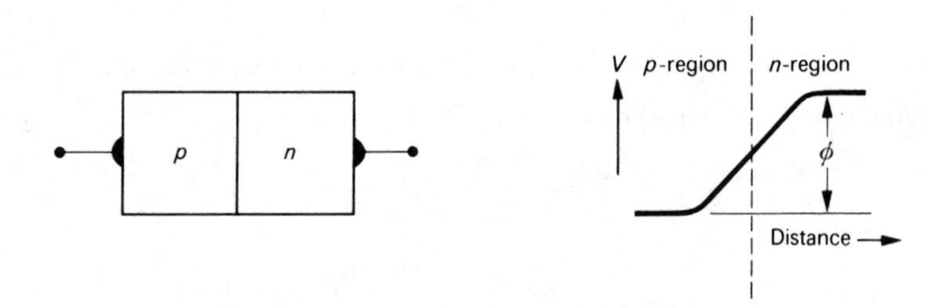

Fig. 26

In Section 3.5 the energy barrier ϕ was obtained as

$$\phi = \frac{kT}{e}\ln\left(\frac{N_a N_d}{n_i^2}\right)$$

where N_a, N_d are the concentrations of the acceptor and donor atoms respectively and n_i is the intrinsic concentration. Here we have $\phi = W$, $n_i = 10^{13}$, $N_a = 10^{17}$, $N_d = 5 \times 10^{16}$ and $kT/e = 0{\cdot}026$ $(T = 300\,\mathrm{K})$.

Hence
$$W = 0{\cdot}026 \times 2{\cdot}3 \times \log_{10}\left(\frac{10^{17} \times 5 \times 10^{16}}{10^{26}}\right)$$
$$= 0{\cdot}026 \times 2{\cdot}3 \times \log_{10}(5 \times 10^{7})$$
$$= 0{\cdot}026 \times 2{\cdot}3 \times 7{\cdot}699$$
or
$$W = 0{\cdot}46\,\mathrm{eV}$$

In Appendix C it is shown that the reverse saturation current density J_s is given by

$$J_s = e\left[\frac{D_p p_n}{L_p} + \frac{D_n n_p}{L_n}\right]$$

where n_p, p_n are the minority carrier densities in p-type and n-type material respectively, while L_p, L_n are the corresponding diffusion lengths. Here we have $D_p = 50\,\mathrm{cm^2\,s^{-1}}$, $D_n = 100\,\mathrm{cm^2\,s^{-1}}$ and $L_p = L_n = 0{\cdot}08\,\mathrm{cm}$.

Also, if N_d, N_a are the donor and acceptor concentrations and n_i is the intrinsic concentration we have

$$p_n = n_i^2/N_d = 10^{26}/(5 \times 10^{16}) = 0{\cdot}2 \times 10^{10}$$
$$n_p = n_i^2/N_a = 10^{26}/10^{17} = 10^{9}$$
with
$$J_s = \frac{1{\cdot}6 \times 10^{-19}}{0{\cdot}08}(50 \times 0{\cdot}2 \times 10^{10} + 100 \times 10^{9})$$
$$= \frac{1{\cdot}6 \times 10^{-19} \times 2 \times 10^{11}}{0{\cdot}08}$$
or
$$J_s = 0{\cdot}4 \times 10^{-6}\,\mathrm{A\,cm^{-2}} = 0{\cdot}4 \times 10^{-2}\,\mathrm{A\,m^{-2}}$$

4

Solid-state devices

The development of various semiconductor materials such as silicon and the improvement in manufacturing techniques, has led to the construction of a wide variety of solid-state devices. The most important of these devices will be considered now, in terms of their main characteristics and applications in the field of electronics.

4.1 Semiconductor diodes

Various diodes now in use are the junction diode, Zener diode, varactor diode and tunnel diode. Some special diodes such as the Gunn diode, Impatt and Trapatt diodes are described in Appendix F, while other semiconductor diodes like the photo-diode and light-emitting diode are described in Chapter 6.

Junction diode

An ordinary *p-n* junction diode can be used for rectifying purposes because it conducts mainly in one direction only. To determine its current–voltage characteristic, consider a *p-n* junction diode connected first with *forward bias* as shown in Fig. 27(a) and then with *reverse bias* as shown in Fig. 27(b).

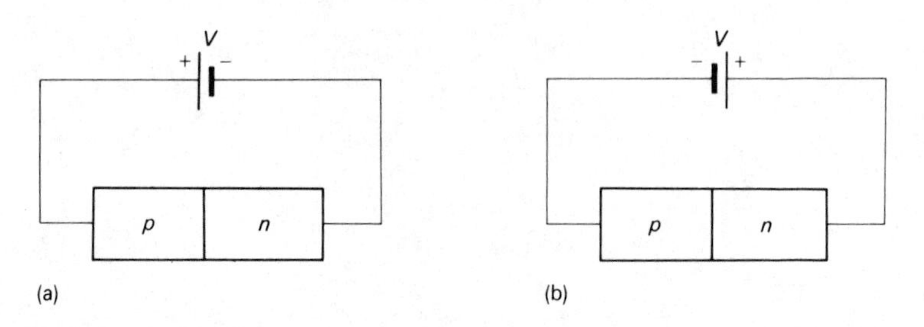

Fig. 27

With forward bias applied the contact potential across the junction is reduced as in Fig. 28(b). Holes in the *p*-type material move easily into the *n*-type material and electrons in the *n*-type material move easily into the *p*-type material. This produces a large current flow through the diode which increases as the applied voltage increases.

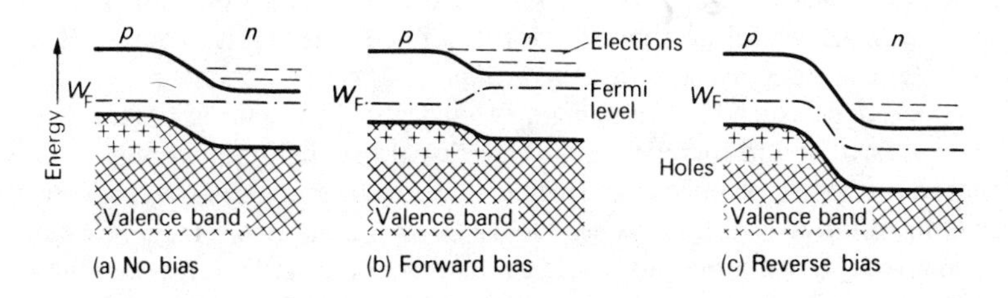

Fig. 28

When reverse bias is applied, the contact potential is increased as shown in Fig. 28(c) and the majority carriers are unable to cross the junction. Only a small number of minority carriers move across the junction due to their thermal energy and produce a small leakage current through the device. In Appendix D it is shown that the characteristic equation of current flow through the device is given by

$$I = I_s(e^{eV/kT} - 1)$$

where I_s is the reverse saturation current and V is the applied voltage. A typical plot of diode current I is shown in Fig. 29.

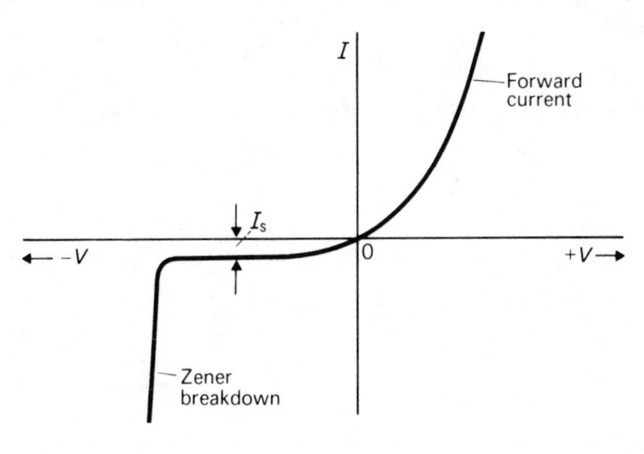

Fig. 29

Due to its rectifying property, large *p-n* diodes are used for converting a.c. power into d.c. power, and at low power levels junction diodes are used as detectors in communication systems.

Zener diode

When a *p-n* junction is reverse biassed there is a large increase in current at a particular value of voltage. This is known as a *breakdown* phenomenon and there are two mechanisms causing this breakdown. The first is called the *Zener effect*[12] and the second is called the *avalanche effect*.[13]

In *p-n* diodes which have thin, highly doped, abrupt junctions, the transition region is narrow and there is a large electric field in the junction region. This causes electrons to be pulled out from the lattice structure and the consequent breaking of the covalent bonds produces a large number of electron-hole pairs, which gives rise to a large current. The phenomenon usually occurs at voltages around 5 V or less and is called the *Zener effect*. An alternative explanation makes use of quantum principles and is called tunnelling. (See the 'tunnel' diode later in this section.)

In ordinary *p-n* junctions at larger voltages, the electrons can also be accelerated to high energies and knock out other electrons from the lattice. These electron-hole pairs in turn produce further collisions and more electron-hole pairs are created. This carrier multiplication process is called the *avalanche effect*. It is illustrated in Fig. 30(a).

Around about 5 V the two effects tend to occur together and diodes which are designed to operate at about 5 V are called Zener diodes. They are used as voltage stabilisers because the voltage across them is remarkably constant over a wide range of current values. The diode also has a very low temperature coefficient and is therefore fairly insensitive to temperature changes. Moreover, the breakdown is not harmful if the diode dissipation is not exceeded. This is ensured by using a series resistor R to limit the current and is shown in Fig. 30(b).

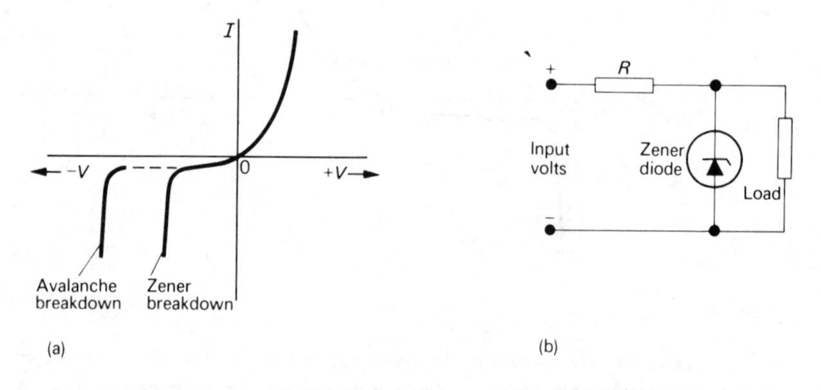

Fig. 30

Varactor diode

A *p-n* junction diode with reverse bias can function as a variable capacitance. Due to the reverse bias, the charge carriers are drawn away from the junction as in Fig. 31(a). The region on either side of the junction is therefore depleted of charge and is called the *depletion layer*. The depletion layer has charges on either side of it and therefore behaves as a parallel plate capacitor.

The thickness of the depletion layer varies with the applied bias and a variable junction capacitor can be produced by varying the reverse bias. For an abrupt *p-n* junction, the variable capacitance C_j of the junction is given by

$$C_j = \frac{C_0}{(V_i + |V|)^{1/2}}$$

where C_0 is a constant depending on the amount of doping and junction area, V_i is the internal junction voltage of about 1 V and $|V|$ is the modulus of the *reverse* bias voltage.

In the case of a graded junction diode, the square root sign is replaced by a cube root sign. Typically C_j varies between 50 pF and 100 pF for an abrupt junction and between 5 pF and 50 pF for a graded junction. A typical characteristic is shown in Fig. 31(b).

The varactor diode finds application in FM modulators for producing an FM signal, or in parametric amplifiers for low-noise amplification.

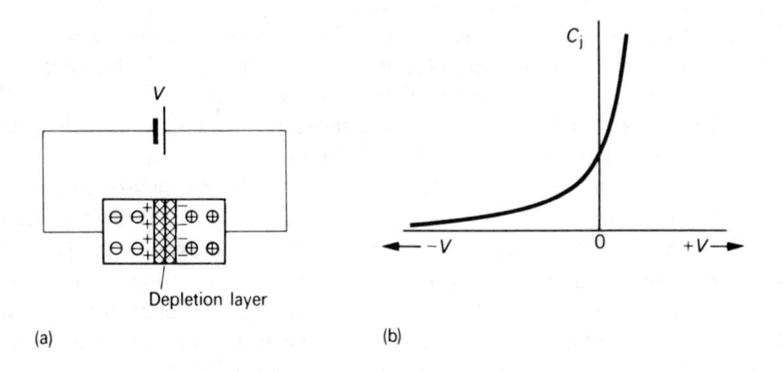

Fig. 31

Tunnel diode[9,14]

If a *p-n* diode is doped heavily in both the *p*-region and *n*-region, the diode has a *negative* resistance over part of its characteristic when forward bias is applied. The negative resistance effect can be explained readily using quantum-mechanical principles.

Since heavy doping produces a very thin barrier region, electrons with wavelike properties can 'tunnel' through the barrier from the *n*-region into the *p*-region with only a small forward bias. Hence, the name *tunnel* diode or *Esaki* diode due to its discovery by Esaki.

However, tunnelling is only possible if electrons in occupied energy states in the *n*-region can tunnel across to unfilled energy states in the *p*-region. For this to be possible, the energy band structure must be as shown in Fig. 32(a). This explains the negative resistance characteristic shown in Fig. 33.

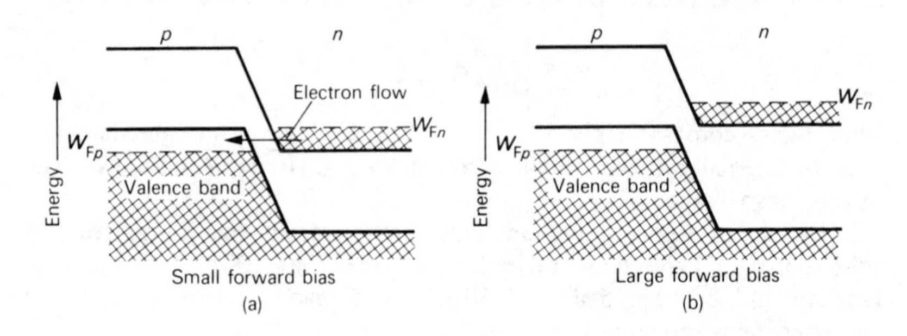

Fig. 32

With a small forward bias applied, occupied energy states of the conduction band in the *n*-region appear opposite unfilled energy states in the valence band. Hence, electrons can tunnel across and the current increases with a small forward bias. As the bias voltage becomes larger, as in Fig. 32(b), the tunnelling current decreases to give the negative resistance part shown in Fig. 33. When the tunnelling current falls to zero, the characteristic rises again as for the normal junction diode characteristic with forward bias.

Due to the negative resistance effect, a tunnel diode can be used to produce oscillations up to several GHz but its operation is critical. Tunnel diode amplifiers are used in satellites such as Intelsat 4. They provide about 10–15 dB gain and have a noise figure around 5 dB. However, as the tunnelling action is very fast, it has an additional use as a switch in logic circuits with operating times around a nanosecond. In this case, the diode switches from stable position P to stable Q or vice versa as shown in Fig. 33.

Example 9

Explain what is meant by the depletion layer in a semiconductor *p-n* junction and sketch the variation of both doping density and space-charge density in the depletion region of an abrupt *p-n* junction.

Derive from first principles the approximate capacitance, for zero applied bias, of an abrupt *p-n* junction that has the following properties.

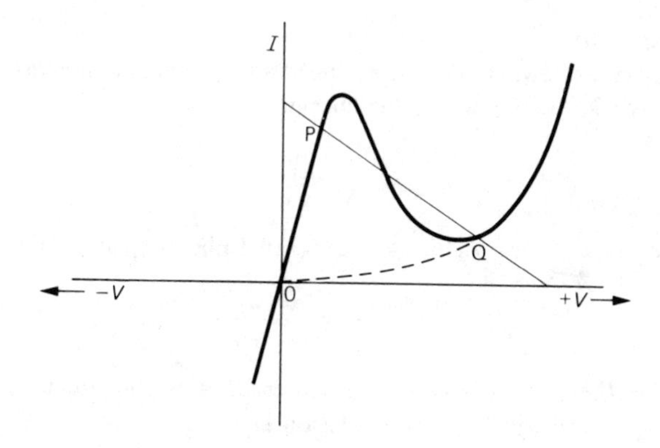

Fig. 33

Acceptor doping density	$N_A = 10^{22}/\text{m}^3$
Donor doping density	$N_D = 10^{19}/\text{m}^3$
Contact potential	$V_0 = 0{\cdot}75\ \text{V}$
Junction area	$A = 10^{-6}\ \text{m}^2$
Relative permittivity of semiconductor material	$\varepsilon_r = 12$

If a reverse bias of 6 V is now applied to the junction find the new value of the depletion layer capacitance. (U.L.)

Solution

When a reverse bias is applied to a *p-n* junction, the charge carriers near the depletion layer are drawn further away from the junction, and the depletion layer at the junction is widened. The depletion layer thickness varies with the applied bias and so it behaves as a variable capacitance. The variation of doping density and space-charge density for an abrupt *p-n* junction are shown in Fig. 34.

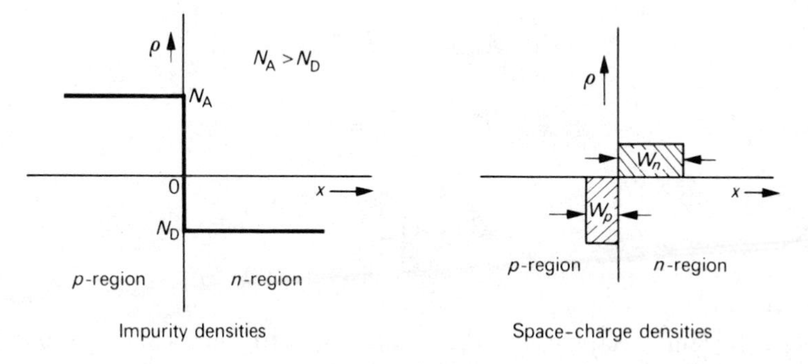

Fig. 34

Junction capacitance

The junction capacitance C_j *decreases* due to an incremental increase in applied *reverse* voltage V. Hence, if q is the charge,

$$C_j = -\frac{dq}{dV}$$

Since the junction also behaves as a parallel plate capacitor we have

$$C_j = \frac{\varepsilon_0 \varepsilon_r A}{w} = \frac{\varepsilon_0 \varepsilon_r A}{(w_p + w_n)}$$

where $\varepsilon_0 \varepsilon_r$ is the permittivity of the material, A is the junction area and $w = (w_p + w_n)$ is the width of the depletion layer.

The material is neutral as a whole and so the charge on either side of the junction must be equal. For acceptor and donor concentrations N_A, N_D respectively we have

$$q = N_A e A w_p = N_D e A w_n$$

or

$$(w_p + w_n) = \frac{q}{eA}\left(\frac{1}{N_A} + \frac{1}{N_D}\right)$$

Substituting in C_j above then yields

$$C_j = \frac{\varepsilon_0 \varepsilon_r e A^2}{q(1/N_A + 1/N_D)} = -\frac{dq}{dV}$$

or

$$q dq = -A^2 C_0 dV$$

where $C_0 = \varepsilon_0 \varepsilon_r e/(1/N_A + 1/N_D)$ is a constant.

For a charge variation 0 to q, let the junction potential vary from the barrier potential V_B to V. Integration then yields

$$q^2/2 = A^2 C_0 (V_B - V)$$

or

$$q = A[2C_0(V_B - V)]^{1/2}$$

with

$$dq/dV = -\tfrac{1}{2}A[2C_0(V_B - V)]^{-1/2}$$

Hence

$$C_j = -\frac{dq}{dV} = A\left[\frac{C_0}{2(V_B - V)}\right]^{1/2}$$

or

$$C_j = A\left[\frac{C_0}{2V_B}\right]^{1/2} \quad \text{(when } V = 0\text{)}$$

Problem

$N_A = 10^{22}/\text{m}^3$, $N_D = 10^{19}/\text{m}^3$, $\varepsilon_r = 12$, $A = 10^{-6}\text{m}^2$, and $V_B = 0{\cdot}75$ V.

Here $$C_0 = \frac{\varepsilon_0 \varepsilon_r e}{(1/N_A + 1/N_D)} = \frac{8{\cdot}85 \times 10^{-12} \times 12 \times 1{\cdot}6 \times 10^{-19}}{(10^{-22} + 10^{-19})}$$

$$= \frac{8{\cdot}85 \times 12 \times 1{\cdot}6 \times 10^{-31}}{10^{-19}}$$

or $$C_0 = 170 \times 10^{-12} = 1{\cdot}7 \times 10^{-10}\ \text{F}$$

Hence $$C_j = 10^{-6}\left[\frac{1{\cdot}7 \times 10^{-10}}{2 \times 0{\cdot}75}\right]^{1/2} = 10^{-6} \times 1{\cdot}06 \times 10^{-5}$$

or $$C_j = 10{\cdot}6\ \text{pF}$$

With the *reverse* bias $V = -6$ V we obtain

$$C_j = 10^{-6}\left[\frac{1{\cdot}7 \times 10^{-10}}{2 \times 6{\cdot}75}\right]^{1/2} = 10^{-6} \times 3{\cdot}54 \times 10^{-6}$$

or $$C_j = 3{\cdot}54\ \text{pF}$$

4.2 Junction transistor[2,10]

The two types of junction transistors used are the *npn* transistor or the *pnp* transistor. The principle of operation of both is very similar, but the *npn* transistor is more widely used because it requires a positive voltage supply which is convenient. A schematic arrangement of both types of transistor is shown in Fig. 35 where the three terminals to each transistor are known as the *emitter*, *base* and *collector*.

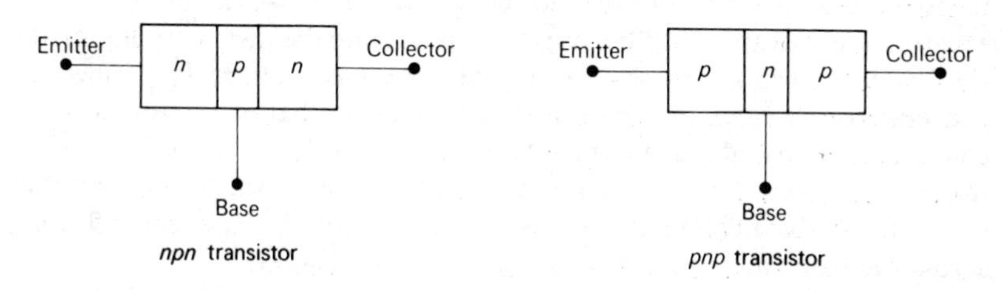

Fig. 35

In the *npn* transistor, a region of *p*-type material lies between two regions of *n*-type material thus forming two *p-n* junctions back to back. In the *pnp* transistor, a region of *n*-type material lies between two regions of *p*-type material and it also forms two *p-n* junctions back to back. The *p-n* junctions can be formed in various ways such as the grown junction, the fused or alloyed junction and the planar or diffused junction.

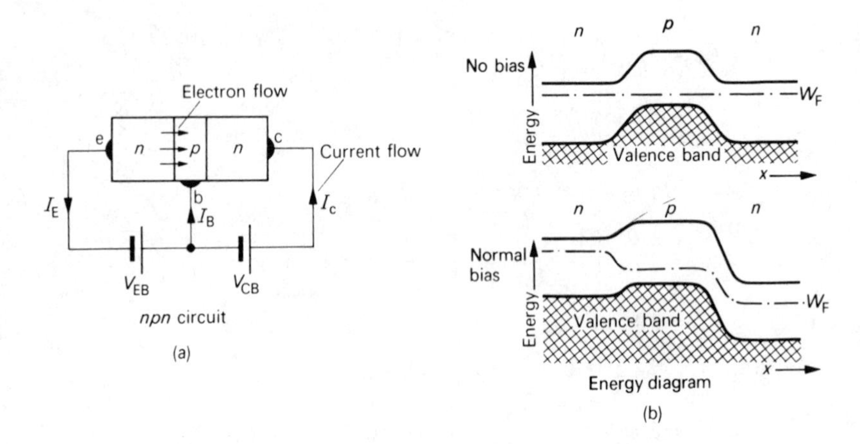

Fig. 36

The operation of an *npn* transistor can be seen from Fig. 36(a) which shows the biassing arrangement for the device. In this arrangement, the *p-n* junction on the left is forward biassed while the *p-n* junction on the right is reverse biassed. The majority carriers in the *n*-type region on the left are electrons which are repelled by the negative pole of the left-hand battery and flow into the base region. Since the base region is kept thin, only a few electrons recombine with holes in the base and are lost. Most of them diffuse across the base region into the collector due to the forward biassing of the right-hand *p-n* junction.

In terms of the energy diagram shown in Fig. 36(b), the potential barrier of the forward biassed *p-n* junction is lowered by the emitter-base voltage V_{EB} and electrons are thus able to diffuse over the top of it into the energy 'trough' on the right which is created by the positive collector-base voltage V_{CB}. Since the conventional direction of current flow is opposite to that of electron flow, the collector current I_C flows into the collector region as shown in Fig. 36(a), while the emitter current I_E flows out from the emitter region. A small base current I_B flows into the base region and is due to some holes in the base region flowing across the base-emitter junction which is forward biassed.

In order to ensure that many electrons cross into the base region, the emitter region is heavily doped, while the base region is only lightly doped and is kept thin to reduce recombination. The collector region is also lightly doped in order to reduce the collector-base capacitance.

The operation of a *pnp* transistor is very similar, except that holes flow across the base region into the collector and this is also the direction of flow of the collector current. Since the charge carriers in these transistors are either holes or electrons, they are also called *bipolar* transistors. A biassing arrangement and energy diagram for the *pnp* transistor is shown in Fig. 37.

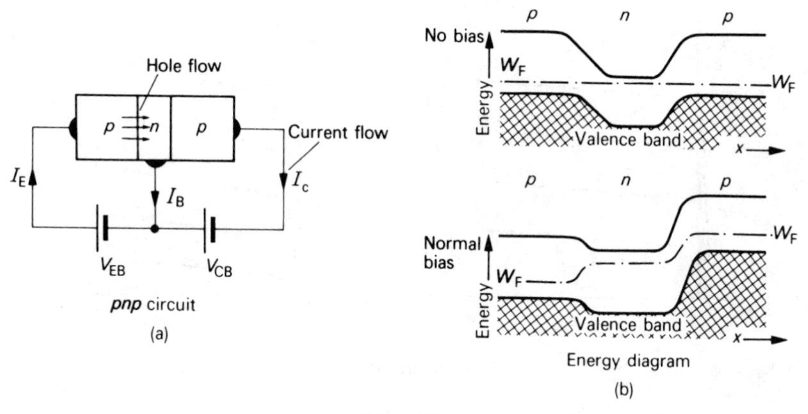

Fig. 37

The bipolar transistor can be operated with three different circuit arrangements generally known as the common-emitter, common-base and common-collector configurations. They are shown in Fig. 38.

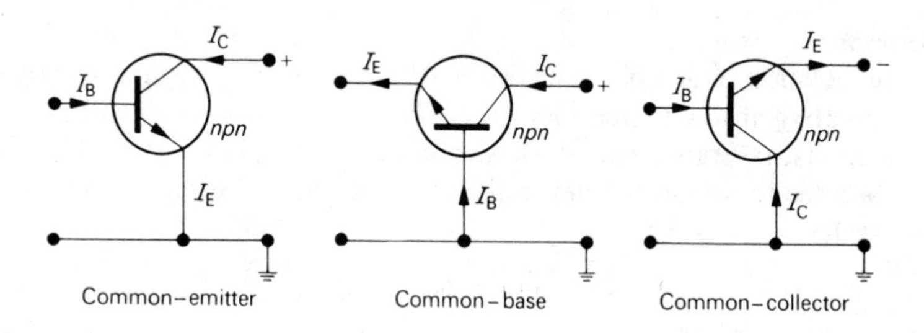

Fig. 38

The two most useful arrangements are the common-base and common-emitter configurations which yield respectively the two types of characteristic curves shown in Fig. 39.

Common-base configuration

The basic current equation for the *npn* transistor is obtained using Kirchhoff's current law, for which the positive direction is assumed for current flowing *into* the device. Hence we obtain from Fig. 38

$$I_C + I_B - I_E = 0$$

or

$$I_E = I_C + I_B$$

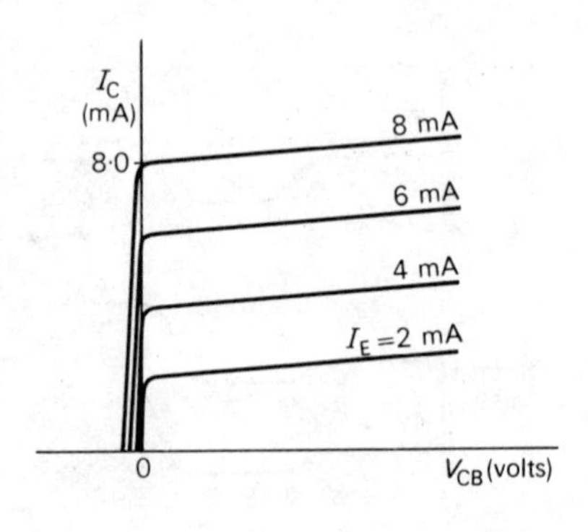

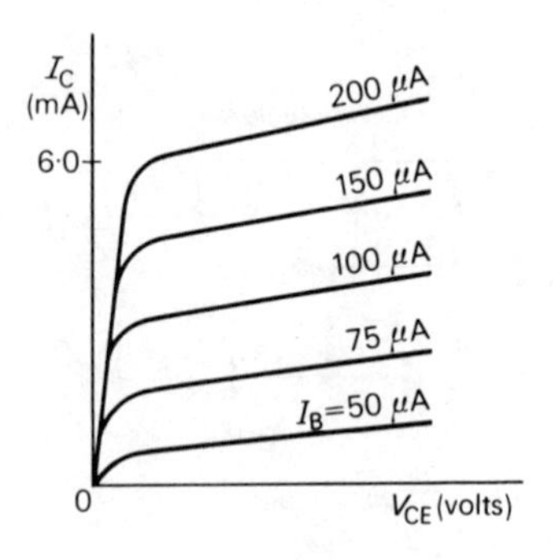

Fig. 39

Since the collector current is not exactly equal to the emitter current due to recombination, the current amplification factor α is defined by

$$\alpha = I_C/I_E$$

or

$$I_C = \alpha I_E$$

Comments

1. In practice, α is usually between 0·95 and 0·99 in typical transistors operating at low frequencies.
2. A reverse saturation current or leakage current I_{CBO} flows from collector to base due to minority holes in the collector region. Hence, we have more exactly

$$I_C = \alpha I_E + I_{CBO}$$

3. I_{CBO} is about 10^{-8} A in silicon transistors and can usually be neglected.

Common-emitter configuration

From Fig. 38, the common current equation for the *npn* transistor is given by

$$-I_E + I_B + I_C = 0$$

or

$$I_E = I_B + I_C$$

Here, it is more useful to define a current amplification factor β such that

$$\beta = I_C/I_B$$

or

$$I_C = \beta I_B$$

Using $I_C \simeq \alpha I_E$ and substituting for I_E above then yields

$$I_C/\alpha = I_C/\beta + I_C = I_C(1/\beta + 1)$$

with $$1/\alpha = 1/\beta + 1$$

or $$1 = \alpha/\beta + \alpha$$

and $$\beta - \alpha\beta = \alpha$$

or $$\beta = \alpha/(1-\alpha)$$

Comments

1. The value of β is large and is typically around 40.
2. If the leakage current I_{CBO} is considered, we obtain

$$I_C = \alpha I_E + I_{CBO}$$

or $$I_C = \alpha(I_B + I_C) + I_{CBO}$$

Hence $$I_C = \left(\frac{\alpha}{1-\alpha}\right) I_B + \left(\frac{1}{1-\alpha}\right) I_{CBO}$$

or $$I_C = \left(\frac{\alpha}{1-\alpha}\right) I_B + I_{CEO}$$

where $I_{CEO} = \left[\frac{1}{(1-\alpha)}\right] I_{CBO}$ is the 'leakage' current in the common-emitter configuration.

3. The basic equations for the *npn* transistor are the same as for the *pnp* transistor.

Example 10

Explain the mechanism of current flow across the base of a *pnp* transistor operating under normal bias conditions. Sketch energy-level diagrams in illustration.

A certain transistor has the following properties. Emitter efficiency 99 %, base transmission factor 99·5 %, collector-current multiplication factor 100 %. Calculate the collector current if the base current is 20 μA and the collector-base leakage current with open-circuited emitter is 1 μA. (U.L.)

Solution

For the biassing arrangement shown in Fig. 37(a), a large number of holes are attracted across the emitter-base region due to the forward bias of this *p-n* junction. As the base-collector junction is reverse biased, there is only a small leakage current I_{CBO} flowing due to electrons moving into the collector-base region.

As the base region is small compared to the diffusion length for holes, some holes are lost due to recombination but most of them are drawn across to the collector by the collector voltage V_C. Hence, together with I_{CBO}, it produces the total collector current I_C. The energy diagrams for this arrangement are shown in Fig. 37(b).

Problem

The emitter efficiency is defined as $\gamma = I_h/I_E$ where I_h is the hole current entering the emitter-base region and I_E is the total emitter current. The base transmission factor is defined as $\beta = I'_C/I_h$, where I'_C is the hole current entering the base-collector region, and the collector multiplication factor is defined as $M = I_C/I'_C$ where I_C is the collector current. Hence, we observe that the current amplification factor is given by

$$\alpha = I_C/I_E = \beta\gamma M$$

or

$$\alpha = 0{\cdot}99 \times 0{\cdot}99 \times 1{\cdot}0 = 0{\cdot}985$$

For the *pnp* transistor shown in Fig. 37(a) we have

$$I_E - I_B - I_C = 0$$

and

$$I_C = \alpha I_E + I_{CBO}$$

where I_{CBO} is the leakage current. Hence

$$\frac{I_C}{\alpha} - I_C = I_B + \frac{I_{CBO}}{\alpha}$$

or

$$I_C = \left(\frac{\alpha}{1-\alpha}\right) I_B + \frac{I_{CBO}}{1-\alpha}$$

Substituting the values $I_B = 20\,\mu\text{A}$, $I_{CBO} = 1\,\mu\text{A}$ and $\alpha = 0{\cdot}985$ yields

$$I_C = \frac{0{\cdot}985}{0{\cdot}015} \times 20 \times 10^{-6} + \frac{(1 \times 10^{-6})}{0{\cdot}015}$$

or

$$I_C = 1{\cdot}38\,\text{mA}$$

4.3 Field-effect transistor[3, 15]

In the field-effect transistor or FET, the current is due to majority carriers only and so it is also called a *unipolar* transistor. There are two types of field-effect transistors which are known as the junction field-effect transistor or JFET and the insulated-gate field-effect transistor or IGFET.

The principle of operation of the FET is quite different from that of the junction transistor because the *drift* current through the FET is controlled by an electric field and is not a diffusion process as in the bipolar transistor. The JFET and IGFET are shown in Fig. 40 where the terminals of each device are called the *source*, *gate* and *drain*.

Junction field-effect transistor

In one form the device consists of an *n*-type conducting channel between a *p*-type gate and a *p*-type substrate. At either end there is a metal contact, one

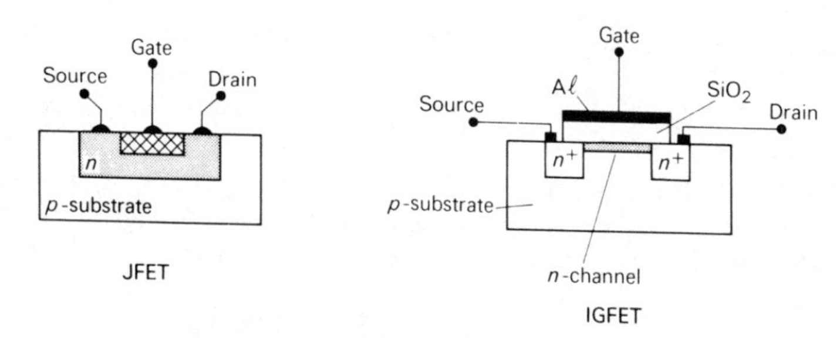

Fig. 40

known as the source and the other known as the drain. The *p*-type gate is formed by diffusion into the *n*-type region, and on application of a negative gate voltage, it behaves as a reverse biassed *p-n* junction. Hence, a depletion region is formed around it which penetrates into the *n*-region. On application of a positive drain voltage, the charge carriers or electrons are drawn across the conducting channel from source to drain. Alternatively, if the channel is *p*-type with outer *n*-type regions, the carriers are holes and the gate voltage is positive.

At low values of V_{DS}, the drain-source voltage, the drain current I_D depends upon V_{DS} and upon V_{GS}, the gate-source voltage. However, as V_{GS} becomes more negative, the depletion region extends deeply into the *n*-region and prevents the current from increasing any further. This is the *pinch-off* or *threshold* voltage V_p, and above V_p, the current is almost independent of V_{DS}.

This mode of operation of the JFET is known as the *depletion mode*, since the current is essentially determined by the extent of the depletion region produced by the negative gate-source voltage. Typical characteristics are shown in Fig. 41. Below the pinch-off voltage V_p, the operation is known as the *triode* region and above V_p, as the *pentode* region, due to similarity with the triode and pentode characteristics respectively.

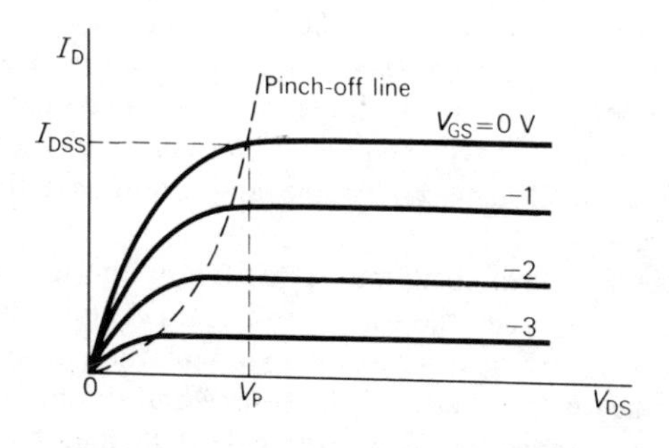

Fig. 41

It is shown in Appendix E that the basic equation of the JFET is given by

$$I_D = \frac{2\sigma ad}{L}\left[V_{DS} - \frac{2}{3}\frac{(V_{DS}+V_{GS})^{3/2}}{V_p^{1/2}} + \frac{2}{3}\frac{V_{GS}^{3/2}}{V_p^{1/2}}\right]$$

where I_D is the drain current, V_{DS} is the drain-source voltage, V_{GS} is the gate-source voltage, V_p is the pinch-off voltage, a is the depletion width at pinch-off, L is the length of the channel, d is the depth of the channel and σ is the conductivity of the channel.

The saturation drain current I_{DS} is obtained when $(V_{DS} + V_{GS}) = V_p$ and is given by

$$I_{DS} = \frac{2\sigma ad}{L}\left[\frac{V_p}{3} - V_{GS} + \frac{2}{3}\frac{V_{GS}^{3/2}}{V_p^{1/2}}\right]$$

and in the particular case when $V_{GS} = 0$ we obtain

$$I_{DS} = I_{DSS} = \frac{2\sigma ad}{L}\frac{V_p}{3}$$

which is illustrated in Fig. 41. An approximate expression for I_{DS} is given by

$$I_{DS} = I_{DSS}\left[1 - \frac{V_{GS}}{V_p}\right]^2$$

and since the mutual conductance $g_m = \partial I_{DS}/\partial V_{GS}$ we obtain

$$g_m = \frac{2I_{DSS}}{V_p}\left[1 - \frac{V_{GS}}{V_p}\right]$$

for an n-channel device with $V_{GS} \leqslant V_p$.

Insulated-gate field-effect transistor[10, 16]

This type of field-effect transistor consists of a metal-oxide semiconductor mounted between a source and drain on a substrate. The device is also known as an MOS field-effect transistor or MOSFET and may be operated in the *enhancement mode* or *depletion mode.* Two possible constructions are the diffused channel MOSFET, in which the conduction channel between source and drain is of the diffused type and the induced channel MOSFET, in which the conduction channel between source and drain is induced. They are illustrated in Fig. 42.

The principle of operation of the MOSFET is due to the presence of a conducting channel between the source and drain. The application of a gate voltage induces an electric field across the channel, which varies its conductivity and hence the current flow between the source and drain.

In the diffused channel MOSFET, the gate is formed by a metal surface insulated by a thin SiO_2 layer from a p-substrate. The source and drain are of

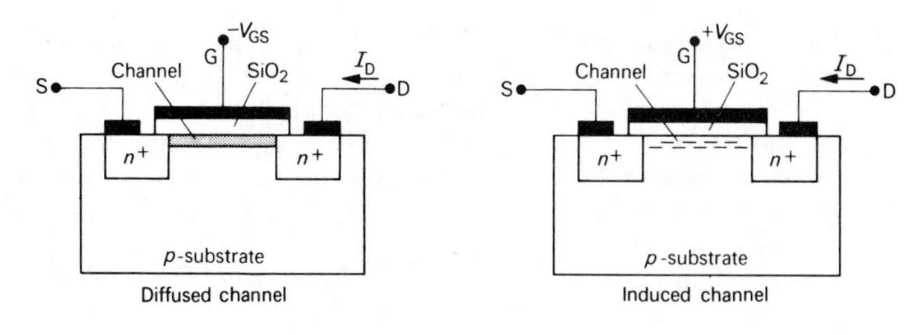

Fig. 42

the diffused n-channel type, which are highly doped regions denoted by n^+, and initially there is a thin conducting channel between them. On application of a negative voltage, the channel conductivity and the current between source and drain is reduced due to depletion of charge carriers, which is dependent on the gate voltage, and this corresponds to the *depletion* mode.

In the induced channel MOSFET, the source and drain are initially insulated from each other. On application of a positive gate voltage, a conducting channel is produced due to an induced *inversion* layer. The conductivity of the layer and the current between the source and drain increase with gate voltage, and this corresponds to the *enhancement* mode.

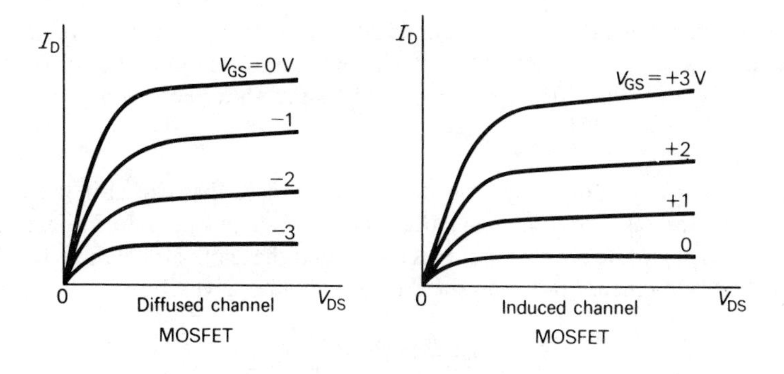

Fig. 43

The characteristics shown in Fig. 43 are similar to that of the JFET. The device has a high input impedance and therefore resembles a pentode in this respect. It is shown in Appendix E that the basic equation of an insulated-gate field-effect transistor (MOSFET) is given by

$$I_D = \frac{\mu \varepsilon_0 \varepsilon_r d}{tL}\left[(V_{GS} - V_p)V_{DS} - \frac{V_{DS}^2}{2}\right]$$

where μ is the effective mobility of electrons for an n-type channel, ε_r is the relative permittivity of the oxide layer and t is its thickness, d is the depth of the channel and L is the channel length. In this expression, the gate-source voltage V_{GS} is assumed positive, while the pinch-off voltage V_p is usually a few volts only. When the drain-source voltage V_{DS} is made equal to $(V_{GS} - V_p)$ the expression becomes

$$I_{DS} = \frac{\mu\varepsilon_0\varepsilon_r d}{tL}\left[\frac{(V_{GS} - V_p)^2}{2}\right]$$

which is known as the *saturation* drain current for a given V_{DS}. Thereafter, it is independent of V_{DS} and corresponds to the levelling off in the current characteristic as is the case with the pentode valve.

Note
Closely associated with MOS technology is the subject of *Integrated Circuits* and recent developments such as the *Charge-Coupled Device* (CCD). Further details of these topics will be found in Appendix H.

Example 11
Describe the principle of operation of an insulated-gate field-effect transistor (MOSFET). Derive the equation

$$I_D = \beta[\tfrac{1}{2}V_D^2 - (V_G - V_p)V_D]$$

(in which the symbols have their usual meaning) and use it to predict the saturation current for a given V_{GS}. It what ways, and why, does the behaviour of actual devices differ from that predicted by the above equation? (C.E.I.)

Solution
The principle of operation of the MOSFET was described earlier in the section and the expression for the drain current I_D is derived in Appendix E. It is similar to that given in the problem if it is assumed that the channel is of the p-type and so the drain current flows out of the device and is given by

$$-I_D = \frac{\mu\varepsilon_0\varepsilon_r d}{tL}[(V_G - V_p)V_D - \tfrac{1}{2}V_D^2]$$

The saturation current I_{DS} occurs when $V_D = (V_G - V_p)$ and hence

$$-I_{DS} = \frac{\mu\varepsilon_0\varepsilon_r d}{tL}\left[\frac{(V_G - V_p)^2}{2}\right]$$

Last part
In actual devices, the saturation drain current increases slightly with drain voltage. This is because the channel length L decreases as the drain voltage increases and so affects I_{DS}. This is particularly true for high-frequency

transistors which are designed to have a short channel length and are therefore affected more by a decrease in L.

Example 12

Referring to simple schematic diagrams, describe the essential constructional features of a metal-oxide-semiconductor type of field-effect transistor, and explain its operation, distinguish between *enhancement* and *depletion*, and discuss the phenomenon of pinch-off.

For such a device give a simple equivalent circuit and outline its physical basis. (U.L.)

Solution

The answer to the first part of the question will be found in Section 4.3.

In the *enhancement* mode the gate voltage is positive and induces a channel between the source and drain. In the *depletion* mode, when a negative gate voltage is used, the device has a narrow channel present between the source and drain.

In the IGFET, the phenomenon of pinch-off occurs when the gate voltage V_{GS} just equals the drain-source voltage for any given V_{GS}. At this value of gate voltage, which is known as the pinch-off voltage V_p, the drain current remains fairly constant and independent of drain voltage V_{DS}. This is because the maximum number of charge carriers induced in the channel are being drawn across to the drain, and increasing drain voltage thereafter does not increase this number further. The phenomenon also occurs in the JFET which was described in Section 4.3.

Equivalent circuit

Since the characteristics of the IGFET given in Fig. 41 are very similar to those of the pentode valve, a simple equivalent circuit for the IGFET resembles that for a pentode and is shown in Fig. 44.

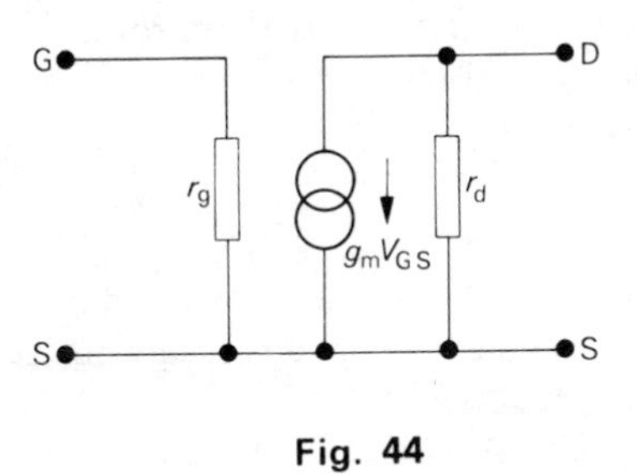

Fig. 44

In Fig. 44, the input circuit gate resistance is r_g which is usually very high and may be neglected. The output circuit consists of a constant current generator $g_m V_{GS}$ shunted by the drain resistance r_d. The transconductance g_m and drain resistance r_d of the device are obtainable from its transfer characteristic and output characteristic respectively. The parameters are defined by the expressions given in Appendix E

$$g_m = \left(\frac{\delta I_D}{\delta V_{GS}}\right)_{V_D \text{ constant}}$$

and

$$r_d = \left(\frac{\delta V_{DS}}{\delta I_D}\right)_{V_{GS} \text{ constant}}$$

4.4 Masers[17]

Masers are devices capable of amplifying weak microwave signals. The word *maser* stands for the expression 'microwave amplification by the stimulated emission of radiation' and masers are usually of the gaseous or solid state forms. A typical gaseous maser is the ammonia maser which is used as an accurate frequency standard and usually called an 'atomic clock'. A well-known solid-state maser is the ruby maser which exists in two forms known respectively as the cavity maser and the travelling wave maser.

Cavity maser

The operation of this solid-state maser depends upon quantum-mechanical principles. Ruby is the aluminium oxide Al_2O_3 which is embedded with some chromium ions. The chromium ion, being paramagnetic, has three unpaired electrons so a magnetic dipole moment is produced, and due to a spin $s = 3/2$, it can be aligned in $(2s+1)$ or four different ways. Hence, in any external magnetic field H_0, the chromium ion has four possible energy states, each associated with a different spin orientation. The four allowed spin orientations and their corresponding energy levels are illustrated in Fig. 45.

Let N_1, N_2 and N_3 be the number of electrons in levels one, two and three respectively. At normal temperatures, the electrons are mainly in the lower energy states as given by the Boltzmann equation

$$\frac{N_1}{N_2} = \frac{e^{-W_1/kT}}{e^{-W_2/kT}}$$

such that $N_1 > N_2 > N_3$. When energy is absorbed by electrons in the lowest level, they are raised to a higher level, and subsequently return to their original level by radiating this energy. This emission of energy takes place spontaneously and is known as *spontaneous emission.*

Hence, transitions with the emission of radiation can only take place if the higher levels are more densely populated than the lower levels. Moreover, these radiative transitions can also be stimulated by using an external source of energy at the correct frequency, and this is known as *stimulated emission.*

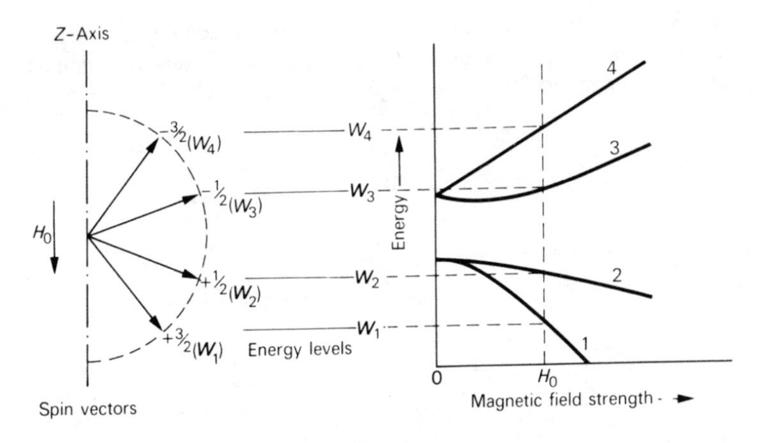

Fig. 45

To produce amplification, electrons must be lifted to energy level W_3 by applying an external RF field such that the pump frequency $f_p = (W_3 - W_1)/h$. The process is known as *population inversion* and the new density of states is such that it gives rise to a 'negative temperature' with $N_3 > N_2 > N_1$ where

$$\frac{N_2}{N_1} = \frac{e^{-W_2/kT}}{e^{-W_1/kT}}$$

or

$$\ln\frac{N_2}{N_1} = -\frac{W_2}{kT} + \frac{W_1}{kT} = \frac{(W_1 - W_2)}{kT}$$

with

$$T = \frac{(W_1 - W_2)}{k \ln (N_2/N_1)}$$

and so T is negative because $W_1 < W_2$.

To obtain operation as a microwave amplifier, a weak external signal is applied at a frequency f_s to stimulate a large number of coherent transitions of electrons from level two to level one, with the consequent emission of radiation according to the equation

$$f_s = \frac{(W_2 - W_1)}{h}$$

and this produces an amplified signal at frequency f_s. However, to prevent incoherent transitions to the lower levels due to thermal collisions and so reduce noise, the ruby crystal must be cooled to a very low temperature with liquid helium or liquid nitrogen.

The construction of a typical cavity maser is shown in Fig. 46. It consists of a microwave resonant cavity which contains a small ruby crystal at one end. The assembly is placed in a flask of liquid helium and mounted between the poles of an electromagnet. The spacing between the energy levels is obtained by adjusting the magnetic field to the required value.

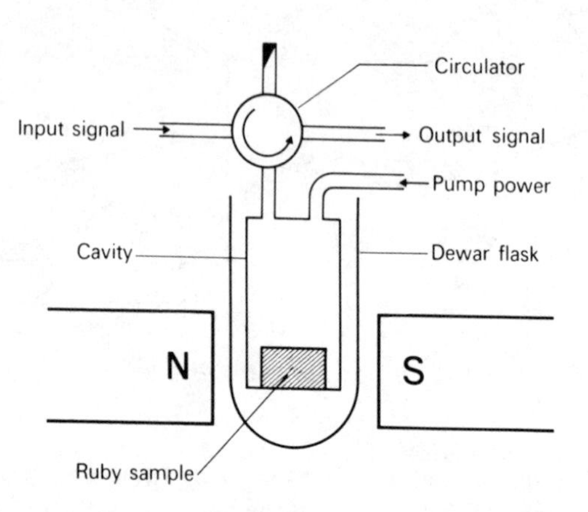

Fig. 46

Signal power from a pump source is coupled to the crystal at a frequency f_p using the resonant mode H_{013}*. To isolate the input from the output, a four-port circulator is used with the input signal coupled in at port one and the amplified output is obtained at port three. The fourth port is terminated in a matched load, to absorb any remaining reflected power.

Masers are often used as low-noise pre-amplifiers in satellite communication systems. A typical C-band maser operating at 4 GHz would have a gain of about 30 dB, a noise temperature around 15 K and a bandwidth of about 10 MHz.

Travelling-wave maser

The main disadvantage of the cavity maser is its narrow bandwidth due to a tightly coupled high Q-circuit. By using a loosely coupled slow-wave structure and a longer sample of material, a wider bandwidth can be obtained without loss of gain. This is illustrated in Fig. 47.

The input signal is loosely coupled to the sample material but interacts with it for a longer time. This is achieved by using the 'comb' slow-wave structure shown in Fig. 47. The active material is 0·05 % chromium in ruby and the input signal is coupled to it to give an amplified signal at the output. The assembly is

* See F. R. Connor, *Wave Transmission*, p. 102, Edward Arnold (1972).

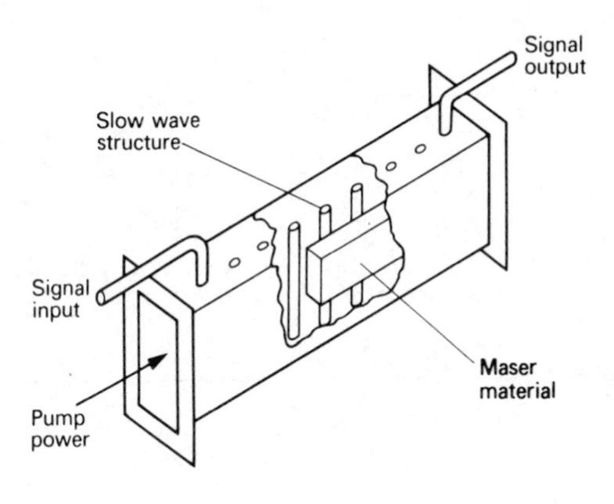

Fig. 47

placed in a short section of waveguide into which is fed RF power at the pump frequency.

A typical device operating at a frequency between 5 and 6 GHz has a forward gain of 20 dB, a 30 dB reverse loss, a 25 MHz bandwidth and a pump frequency of 19 GHz at about 10 mW pump power. A noise temperature around 10 K is attainable using liquid helium, and the device is tunable over its frequency range by means of a magnetic field.

4.5 Lasers[18]

When amplification at optical frequencies is involved, the device is usually called a *laser* which stands for '*l*ight *a*mplification by the *s*timulated *e*mission of *r*adiation'. The laser is capable of producing intense coherent radiation over a narrow band of frequencies in contrast with conventional light sources which produce incoherent radiation over a broad frequency band. As in the case of masers, both spontaneous and stimulated transitions can occur with the emission of radiation.

The most widely used lasers are of the gaseous or solid-state forms. Typical gas lasers are the helium–neon laser which is used as a monochromatic source of light and the CO_2 laser which can provide higher powers for cutting and welding in metallurgical processes. The most common solid-state lasers are the ruby laser and the gallium arsenide (GaAs) laser.

Ruby laser

To obtain stimulated emission at optical frequencies, i.e. laser action, the energy spacing between levels must be large. Ruby with chromium ions has an energy diagram as shown in Fig. 48 and can be used to produce laser action. If green light is used to raise the population in level three, the atoms return to a

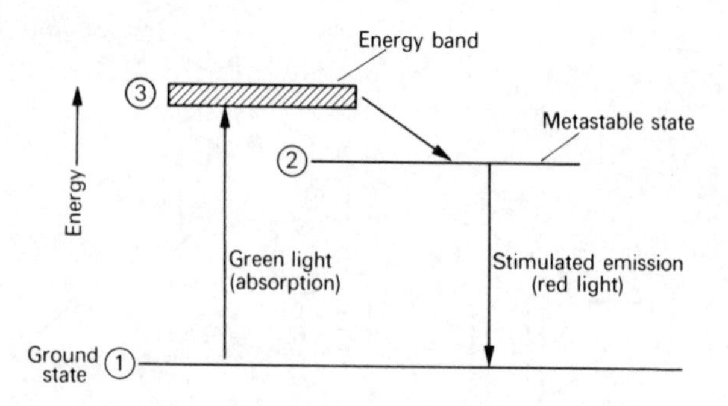

Fig. 48

metastable state at level two very quickly by giving up the energy difference as fluorescence. As the metastable state has a relatively long lifetime, electrons can build up at this level by suitable optical pumping, and population inversion occurs between level two and level one. Electrons in level two finally return to the ground state (level one) by emitting coherent red light.

A typical ruby laser uses a flash tube as the pump source. Two possible geometries are shown in Fig. 49 and comprise either a spherical or elliptical arrangement. Both designs use a pulsed source which causes stimulated emission of photons in various directions. By silvering the end faces, photons along the central axis are reflected backward and forward several times thus

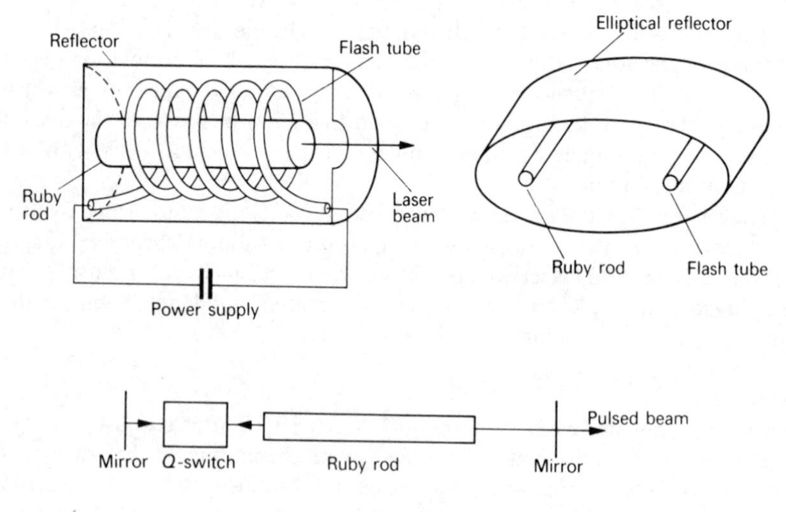

Fig. 49

causing further stimulated emission, until a strongly focussed beam of red light is allowed to escape through a small hole in one end mirror.*

To concentrate light along the central axis, the assembly is surrounded by a reflecting mirror. As only a small portion of the white light energy is used as green light energy in producing laser action, overheating of the crystal is overcome by using a pulsed RF source to flash the tube periodically. However, the output contains a number of small isolated pulses of red light. By means of a '*Q*-switch' the pulses can be made to produce one large pulse of red light. The *Q*-switch may be a mechanically rotating end mirror or an electro-optical shutter which opens periodically. Peak powers of several megawatts and lasting for a few nanoseconds have been obtained using this technique. An alternative means of producing sharp pulses is to use the mode-locking technique, whereby the various oscillating modes of a laser are synchronised for a very short interval of time to produce a large peak pulse.

Gallium arsenide laser

Certain semiconductor materials are capable of producing population inversion when connected to an energy source. A GaAs *p-n* junction can emit incoherent radiation due to the recombination of holes and electrons in the vicinity of the junction. The recombination radiation is due to electrons in the conduction band dropping into 'holes' in the valence band as shown in Fig. 50.

Materials like GaAs are known as direct gap semiconductors, in which the lowest minimum in the conduction band and the highest maximum in the

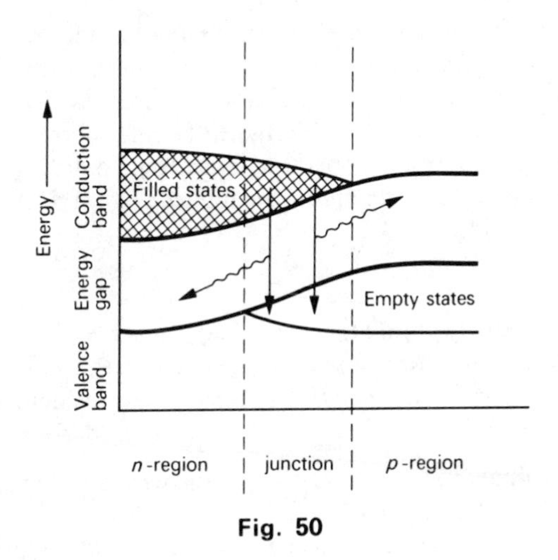

Fig. 50

* A tapered geometry may be used for better focussing.

valence band occur at the same value of momentum. Laser action can only be obtained in a direct gap material as electrons and holes can recombine directly, and most of the energy appears as light. Moreover, the *p*- and *n*-regions have to be heavily doped so that a high density of electron-hole recombinations can take place.

To produce laser action, it is necessary to build up the radiation emitted, especially along the plane of the junction. This can be done by polishing the end faces perpendicular to the plane of the *p-n* junction and pulsing the junction with short pulses of energy as illustrated in Fig. 51.

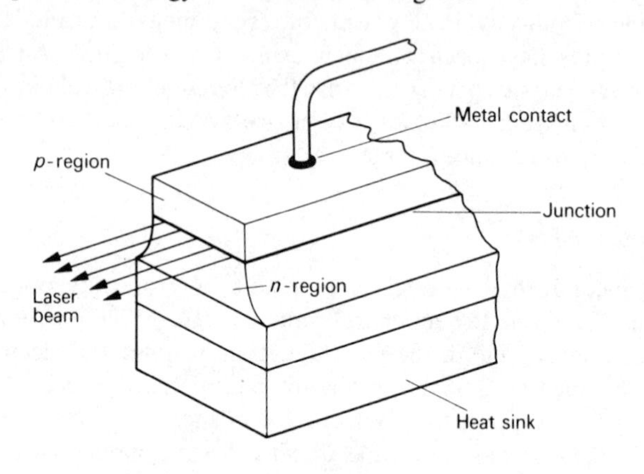

Fig. 51

GaAs lasers are efficient sources of coherent radiation and are noted for their small size. The laser beam may be modulated to convey information and is now being exploited to carry several communication channels by transmission along an optical waveguide in the form of a bundle of quartz fibres. A typical GaAs laser at room temperature has an operating wavelength of 0·9050 μm, produces 6 W of peak radiant power and requires an applied voltage of about 1·5 V. The efficiency is about 40 %.

Example 13

Explain what is meant by *population inversion* of energy levels in a material and show in what sense this effect implies a *negative* temperature.

In a material at 300 K two energy levels have a wavelength separation of 1 μm. Determine

(a) the ratio of upper to lower level occupation densities when the material is in thermal equilibrium;

(b) the *effective* temperature when the levels are equally populated;

(c) the *effective* temperature when the upper level is twice as densely populated as the lower. (Assume unit degeneracy for each level.)

Indicate briefly the relevance of population inversion in the operation of lasers and describe one way by which inversion is achieved. (C.E.I.)

Solution

At normal room temperature, the lower energy level N_1 in a material is more densely populated than a higher energy level N_2 according to the Boltzmann equation (assuming unit degeneracy) as

$$\frac{N_1}{N_2} = \frac{e^{-W_1/kT}}{e^{-W_2/kT}} = e^{(W_2 - W_1)/kT}$$

or

$$T = \frac{(W_2 - W_1)}{k \ln (N_1/N_2)}$$

and since $W_2 > W_1$ and $N_1 > N_2$, the temperature is positive as illustrated in Fig. 52(a).

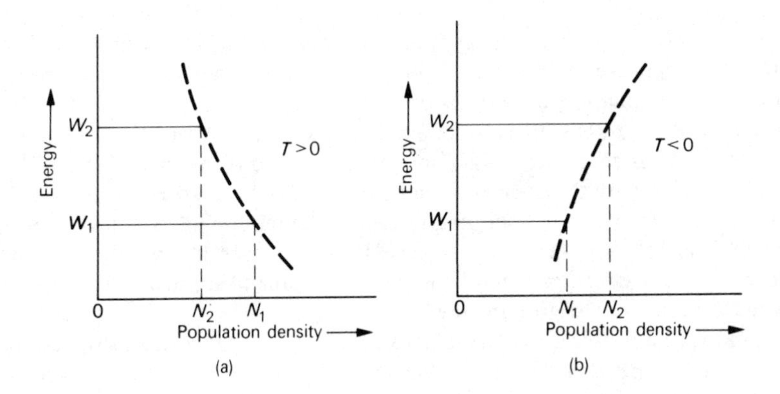

Fig. 52

However, it is possible under certain conditions to make $N_2 > N_1$ with $W_2 > W_1$. This is known as population inversion and from the expression for T above, it will be seen that T has a negative value and the effect is therefore *equivalent* to a negative temperature. This is illustrated in Fig. 52(b).

Problem

(a)

$$\frac{N_2}{N_1} = \frac{e^{-W_2/kT}}{e^{-W_1/kT}}$$

and for transitions between energy levels $\lambda = hc/(W_2 - W_1)$. Hence

$$\frac{(W_2 - W_1)}{kT} = \frac{hc}{\lambda kT} = \frac{6{\cdot}626 \times 10^{-34} \times 3 \times 10^8}{10^{-6} \times 1{\cdot}38 \times 10^{-23} \times 300} = 48{\cdot}01$$

with

$$N_2/N_1 = e^{-48{\cdot}01}$$

and
$$\log_{10}(N_2/N_1) = -0{\cdot}4343 \times 48{\cdot}01 \simeq -21$$
or
$$N_2/N_1 \simeq 10^{-21}$$

(b)
$$N_2/N_1 = e^{-(W_2 - W_1)/kT} = 1$$
or
$$\frac{-(W_2 - W_1)}{kT} = \ln 1 = 0$$
and
$$T \to \pm\infty$$

(c)
$$N_2/N_1 = e^{-(W_2 - W_1)/kT} = 2$$
or
$$-(W_2 - W_1)/kT = 0{\cdot}693$$
with
$$T = \frac{-6{\cdot}626 \times 10^{-34} \times 3 \times 10^{8}}{10^{-6} \times 1{\cdot}38 \times 10^{-23} \times 0{\cdot}693} \simeq -21\,000\ \mathrm{K}$$

Final part

To obtain laser action, the energy spacing between levels must be large, so that stimulated emission occurs in the optical region. For stimulated transitions to occur, the population density in one of the higher energy levels must be increased at the expense of a lower level. This is known as population inversion. A large number of transitions can then be stimulated from the higher level to the lower level with the emission of coherent radiation. A typical material such as ruby has three energy levels and in order to obtain population inversion, a white flash-light source is used to raise the population in level three. Laser action with the emission of red light then occurs due to stimulated transitions from level two to the ground state.

In a four-level laser with a material like neodymium-doped glass, population inversion is more readily obtained than in a three-level laser. Hence, less optical pumping is required to produce laser action between level three and level two.

5

Electron dynamics

The motion of electrons under the influence of electric or magnetic fields plays an important part in the design of various electronic devices. Before considering these devices, it is useful to examine the behaviour of electrons when acted on by electric or magnetic fields.

5.1 Electron motion

A stream of electrons emitted from a metallic surface such as a cathode may be acted on by an electric field, a magnetic field or a combination of both.

Electric field

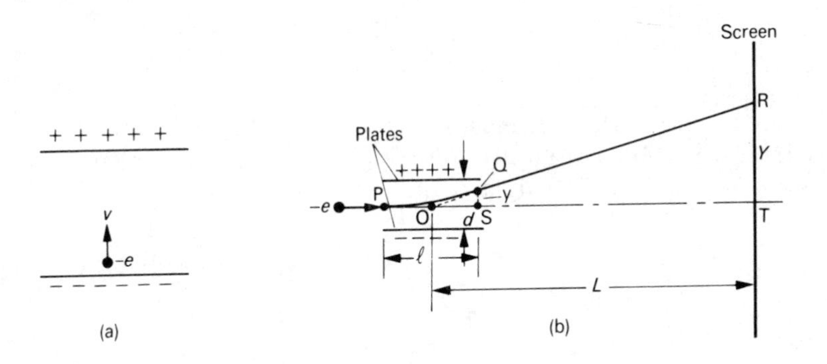

Fig. 53

Consider an electron of mass m and charge e acted on by an electric field E between two parallel plates, as shown in Fig. 53(a). For an electron initially at rest near the negatively charged plate we have

$$eE = ma$$

or

$$a = eE/m$$

where a is the linear acceleration. After a time t, the electron has a velocity v given by

$$v = at$$

or

$$v = eEt/m$$

If the voltage between the plates is V then

$$eV = \tfrac{1}{2}mv^2$$

or

$$v = \sqrt{\frac{2eV}{m}} \quad \text{m s}^{-1}$$

Of special interest is the deflection produced on an electron when travelling between two parallel plates as shown in Fig. 53(b). The electron is assumed to enter the E field normally and follows the curved path PQ, after which it travels along a straight path QR. If the length of the plates is l, the actual path of the electron can be approximated by the linear path OQR, where O is at the centre of the plates, and the electron is assumed to travel with velocity v_x over half the length l. If the component velocities at the edge of the plates are v_x and v_y we have

$$v_x = \sqrt{\frac{2eV_a}{m}}$$

$$v_y = \sqrt{\frac{2eV}{md}}$$

where V_a is the accelerating anode voltage, V is the voltage across the plates and d is the separation of the plates. Hence, we obtain from similar triangles OSQ and OTR, where y is the distance SQ

$$\frac{v_y}{v_x} = \frac{y}{l/2} = \frac{Y}{L}$$

or

$$\sqrt{\frac{V}{dV_a}} = \frac{y}{l/2} = \frac{Y}{L}$$

with

$$\frac{Y}{L}\frac{y}{l/2} = \frac{V}{dV_a}$$

or

$$Y = \frac{L}{2}\frac{V}{d}\frac{l}{V_a} \quad \text{metres}$$

Magnetic field

For an electron entering a magnetic field B with velocity v, the path is a circle

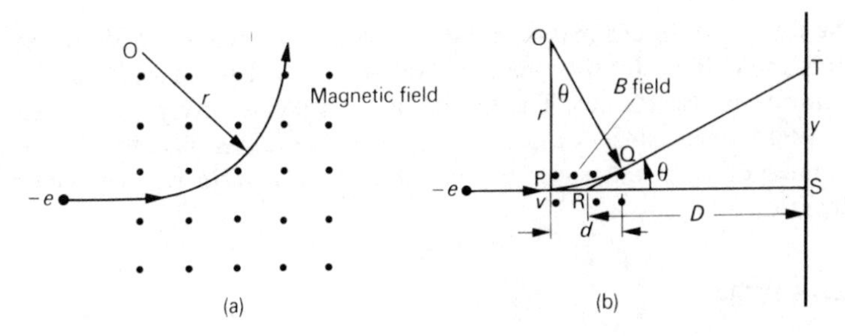

Fig. 54

with radius r. The centrifugal force mv^2/r is balanced by the magnetic force Bev. Hence

$$\frac{mv^2}{r} = Bev$$

or

$$v = \frac{Ber}{m} \qquad \mathrm{m\,s^{-1}}$$

with

$$\omega = \frac{v}{r} = \frac{Be}{m} \qquad \mathrm{rad\,s^{-1}}$$

The magnetic deflection produced by a short magnetic field is illustrated in Fig. 54(b). It is of interest in an important application such as the television tube. For small angles θ we obtain from the similar triangles OPQ and RST

$$\theta = d/r = y/D$$

or

$$y = dD/r = dD\,Be/mv \quad \text{metres}$$

If the velocity v of the electron is due to an accelerating voltage V, we have

$$v = \sqrt{\frac{2eV}{m}}$$

with

$$y = dD\frac{Be}{mv} = \frac{0{\cdot}3dDB}{\sqrt{V}} \times 10^6 \quad \text{metres}$$

using a value of $e/m = 0{\cdot}176 \times 10^{12}\,\mathrm{C\,kg^{-1}}$.

Another application of the magnetic type of deflection is in the mass spectrometer which analyses ions of slightly different mass. By means of a set of slits, ions having a particular velocity range are accelerated through them by an electric field. The ions then enter a uniform magnetic field and due to their slight mass difference describe semicircles of different radii according to the basic equation

$$r = \frac{mv}{Be}$$

The ions of different mass can be collected separately or made to fall on photographic film. In this manner isotopes may be separated and their concentrations ascertained. The technique is used extensively in the chemical field. Alternatively, since the ratio e/m appears in the formula above, knowing the values of B, v and r for a given ion, its e/m ratio can be accurately determined.

Crossed fields

Some electronic devices such as the magnetron depend on the combined action of an electric field and a magnetic field, and are generally called crossed field devices. The behaviour of an electron in a crossed field is shown in Fig. 55(a).

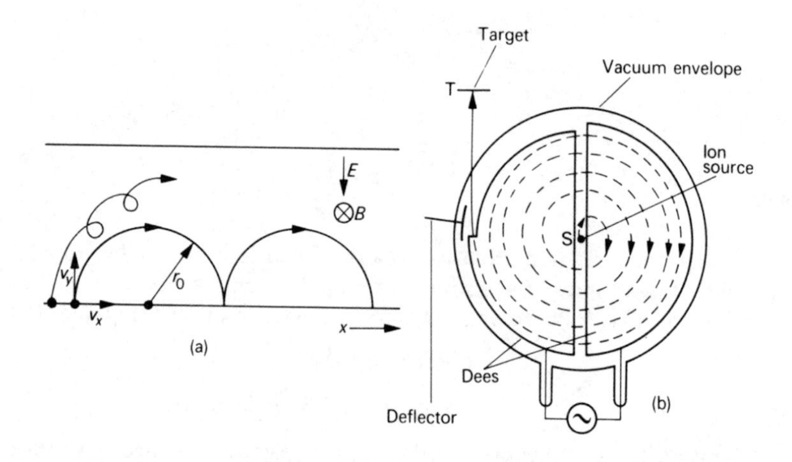

Fig. 55

For perpendicular electric and magnetic fields, the path traced out by the electron depends on its initial velocity and direction. For electrons initially at rest, the paths all lie in a plane and motion is in the general direction of x.

The basic equations of motion are determined by assuming that the electron has component velocities v_x, v_y and v_z. Since the E and B fields are respectively perpendicular to and parallel to v_z, there is no effect on v_z and only the components v_x and v_y are affected by the E and B fields. Hence, we have

$$\frac{dv_x}{dt} = \frac{Bev_y}{m}$$

$$\frac{dv_y}{dt} = \frac{(eE - Bev_x)}{m}$$

which on integration yields

$$v_x = \frac{(eE - m\omega A \cos \omega t)}{Be}$$

$$v_y = A \sin \omega t$$

with
$$v = \sqrt{v_x^2 + v_y^2}$$

where A is an arbitrary constant, $\omega = Be/m$ and $v_x = v_y = 0$ at $t = 0$. The expressions for v_x and v_y yield different electron paths which may be cycloidal or trochoidal depending on the values of E and B.

An important application of crossed fields is the cyclotron illustrated in Fig. 55(b). It is used extensively as a device for accelerating ions to the high energies which are required in physical investigations of the structure of nuclei or for producing radioactive substances.

Essentially, it consists of two metal 'dees' (hollow half-cylinders) in the centre of which is an ion source S. Ions are accelerated across the gap by an RF field applied to the dees. Due to a vertical magnetic field B, the ions follow semicircular paths within the dees and are accelerated to a higher energy at every gap crossing. The ions are finally deflected out by an electric field E to hit a target T. The whole system is enclosed in a vacuum to prevent the ions being scattered by collisions with air molecules. The basic equations of motion are

$$r = mv/Be$$

$$\omega = v/r = Be/m$$

where ω is the cyclotron angular frequency.

Relativistic effects

According to the Special Theory of Relativity, particles accelerated to very high velocities comparable to that of c, the velocity of light, must undergo a mass increase given by

$$m = \frac{m_0}{\sqrt{1 - v^2/c^2}}$$

where m_0 is the rest mass of the particle and m is its relativistic mass at velocity v.

Hence, as v increases, the mass m tends to increase also and consequently, to satisfy the cyclotron equation $r = mv/Be$, the magnetic field B must be increased to keep the ions in step. This increase is achieved in the synchrotron where the magnetic field is varied in synchronism with the mass increase, thus keeping the ions in step with the RF field.

Example 14

Use the basic equation of motion $\mathrm{d}(mv)/\mathrm{d}t = \text{force}$ and the mass–energy equation $W = mc^2$ to establish the relativistic mass equation

$$m = \frac{m_0}{\sqrt{1 - v^2/c^2}}$$

Hence, or otherwise, obtain an expression for v/c for a particle with kinetic energy T and rest energy E_0. Sketch a curve of v/c against T/E_0 emphasising essential features.

Comment briefly on the significance of this curve with regard to particle accelerators. (C.E.I.)

Solution

Since $F = \mathrm{d}(mv)/\mathrm{d}t$, the work done $\mathrm{d}W$ in moving a force F through a distance $\mathrm{d}x$ is given by

$$\mathrm{d}W = F\mathrm{d}x = \mathrm{d}(mv)/\mathrm{d}t \times \mathrm{d}x$$

From the mass–energy relation $W = mc^2$, the change of energy $\mathrm{d}W$ is due to a mass change $\mathrm{d}m$ such that $\mathrm{d}W = c^2\,\mathrm{d}m$ and so we obtain

$$\frac{\mathrm{d}}{\mathrm{d}t}(mv)\,\mathrm{d}x = c^2\,\mathrm{d}m$$

or

$$\frac{m\mathrm{d}v}{\mathrm{d}t}\mathrm{d}x + \frac{v\mathrm{d}x}{\mathrm{d}t}\mathrm{d}m = c^2\,\mathrm{d}m$$

with

$$\frac{m\mathrm{d}x}{\mathrm{d}t}\mathrm{d}v + \frac{v\mathrm{d}x}{\mathrm{d}t}\mathrm{d}m = c^2\,\mathrm{d}m$$

and

$$mv\mathrm{d}v + v^2\mathrm{d}m = c^2\,\mathrm{d}m$$

or

$$\frac{\mathrm{d}m}{m} = \frac{v\mathrm{d}v}{(c^2 - v^2)}$$

After integration and substitution of $\ln A$ for the constant yields

$$\ln m = \ln\frac{1}{\sqrt{(c^2 - v^2)}} + \ln A$$

and since $m = m_0$ when $v = 0$ we obtain

$$\ln m_0 c = \ln A$$

with

$$\ln m = \ln\frac{1}{\sqrt{(c^2 - v^2)}} + \ln m_0 c$$

or

$$m = \frac{m_0}{\sqrt{(1 - v^2/c^2)}}$$

Also, we have

$$mc^2 = \text{Kinetic energy} + E_0$$

or
$$\frac{m_0c^2}{\sqrt{(1 - v^2/c^2)}} = T + E_0 \qquad (E_0 = m_0c^2)$$

with
$$\frac{T}{E_0} = \frac{1}{\sqrt{(1 - v^2/c^2)}} - 1$$

A plot of T/E_0 for various values of v/c is shown in Fig. 56.

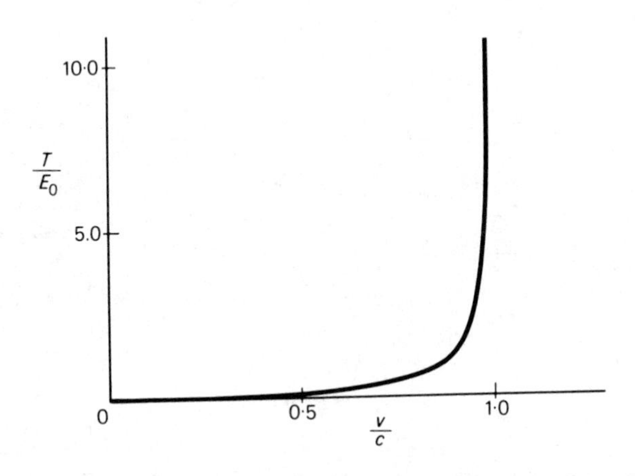

Fig. 56

Comments
For velocities up to about 0·5 c, little kinetic energy is required, but to increase the velocity higher requires a large amount of kinetic energy. This is because most of the extra energy is used to increase the mass of the particle. Hence, accelerators use protons rather than electrons as the latter attain their relativistic mass at relatively low velocities.

Furthermore, to overcome the relativistic mass increase, a synchrocyclotron varies its frequency to keep in step as m increases, while the synchrotron increases the magnetic field B linearly with m to keep the orbital radius constant.

5.2 Electron focussing

A beam of electrons which is diverging from a point source may be re-focussed to a point using an electrostatic or magnetic 'lens'. In Fig. 57, a beam of electrons is seen diverging from a point along the central axis. Electrostatic focussing can be achieved by allowing the diverging beam to enter a nonuniform electric field. The field may be produced by using two coaxial cylinders of equal or unequal radii, which are given different potentials V_1 and V_2 as shown in Fig. 57(a) and Fig. 57(b) respectively.

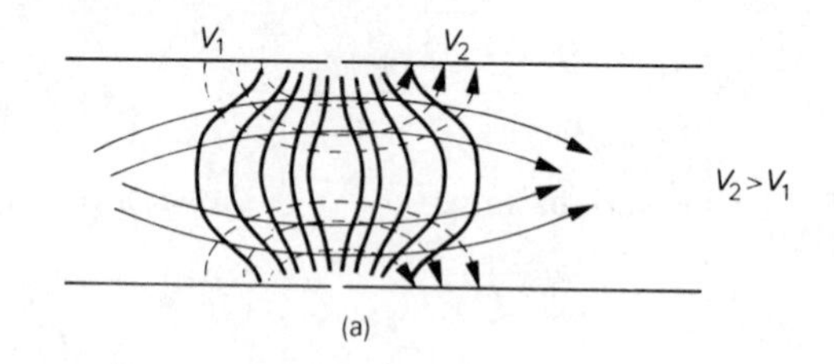

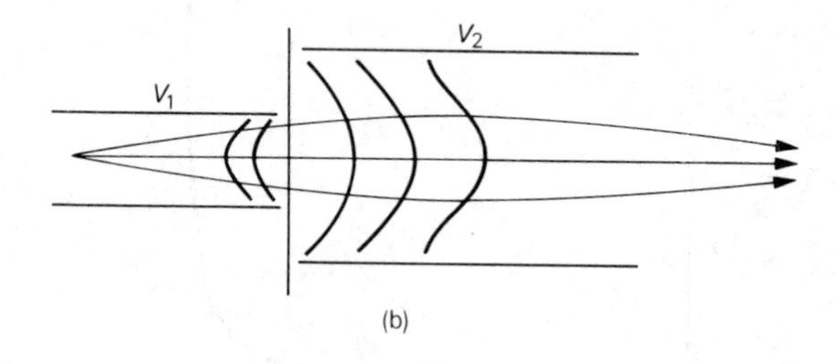

Fig. 57

The field lines and equipotentials are also shown in Fig. 57(a) and the diverging electrons are bent back towards the axis, in a manner similar to that of light in passing through a focussing lens. A detailed analysis of the electron paths leads to a study of electron optics.[19]

In magnetic focussing, a short solenoid or magnet may be used as shown in Fig. 58. In Fig. 58(a), the perpendicular components of the magnetic field cause the diverging electrons to spiral back to a point on the axis and are thus re-focussed. In Fig. 58(b), a diverging electron leaving the axis at angle α is likewise made to spiral back and cross the axis at a point further down the solenoidal axis. The period of rotation of the electrons is $T = 2\pi m/Be$, if m/e is assumed constant, and so all the electrons have the same period of travel and reach the focussing point P together again after the same time T.

Both the above methods, however, are prone to limitations as in the case of an optical lens. These are usually spherical aberration and distortion, and in addition, the magnetic lens produces some astigmatism. By proper design however, these effects can be made reasonably small.

5.3 Electron emission

The emission of electrons from a metallic surface is basic to many electronic

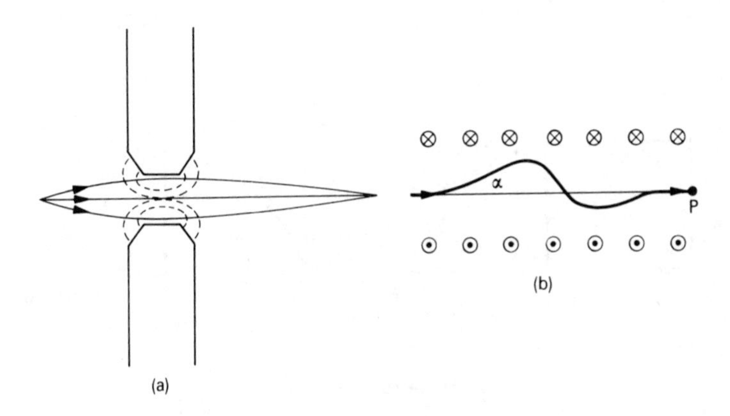

Fig. 58

devices. Four methods may be employed and are known as thermionic emission, photoelectric emission, secondary emission and field emission.

Thermionic emission

When a metallic conductor is heated, the 'free' electrons move about with greater energy in the conductor. Hence, if sufficient heat is applied, they can acquire enough kinetic energy to leave the conductor surface, and this phenomenon is called thermionic emission.

In the conductor, the free electrons have energies which correspond to a Fermi-Dirac distribution function $P(w)$ given by

$$P(w) = \frac{1}{1 + e^{(W - W_F)/kT}}$$

where k is Boltzmann's constant, T is the absolute temperature of the conductor and W_F is the Fermi level of the electrons in the metal. A plot of $P(W)$ is shown in Fig. 59(a) for various temperatures T.

In Fig. 59(a), $P(W)$ represents the probability that an electron has energy W below W_F. At $T = 0\,\mathrm{K}$ this probability is unity as all the electrons occupy the lower energy states. Furthermore, when $T > 0\,\mathrm{K}$, the Fermi level W_F represents an energy level of probability 0·5. As the temperature is increased, some electrons have a finite probability of having energies greater than W_F as seen from Fig. 59(a).

The energy density function $\mathrm{d}N/\mathrm{d}W$ which represents the number of electrons, per unit volume, with energies between W and $(W + \mathrm{d}W)$ becomes

$$\frac{\mathrm{d}N}{\mathrm{d}W} = \frac{AW^{1/2}}{1 + e^{(W - W_F)/kT}}$$

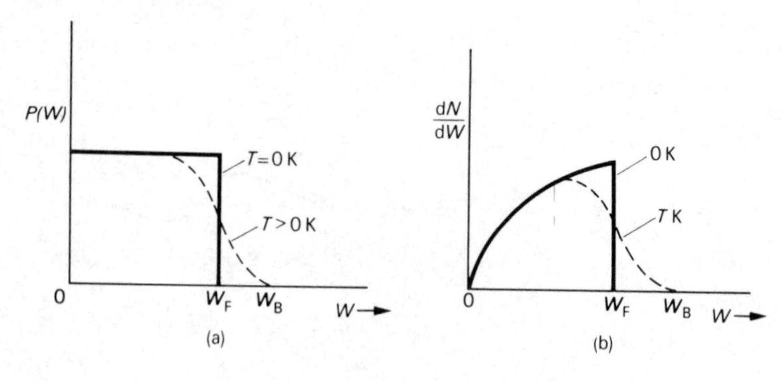

Fig. 59

where $A = (128\pi^2 m^3)^{1/2}/h^3$. The function is plotted in Fig. 59(b) and it will be observed that at $T = 0$ K the lower energy states are occupied up to a maximum value of $W = W_F$.

Hence, the electrons have finite energy and not zero energy even at absolute zero temperature. As the temperature is increased, the energy states are changed slightly to give a maximum value of W_B. The total number N of electrons with energy W such that $0 < W < W_F$ is given by

$$N = \int_0^{W_F} AW^{1/2}\,\mathrm{d}W = \tfrac{2}{3}A\left[W^{3/2}\right]_0^{W_F}$$

or

$$N = \tfrac{2}{3}AW_F^{3/2} = \frac{8\pi(2mW_F)^{3/2}}{3h^3}$$

as shown in Appendix C.

Since some electrons occupy energy states up to W_B, W_B is called the *surface barrier energy* which must be overcome for an electron to leave the metallic surface. Hence, the energy W_f required is given by

$$W_f = W_B - W_F$$

where W_f is called the work function of the material and is expressed in electron volts. Typical values are given in Table 3.

Table 3

Material	W_f (eV)	*Operating temperature* (K)
Tungsten	4·5	2500
Tantalum	4·1	2400
Thoriated-tungsten	2·6	2000
Oxide-coated	1·0	1100

To evaluate the saturation current density J obtainable from an emitter, it is necessary to know the number of electrons leaving *normal* to the surface with energies $W > W_B$. The expression for J is known as the Richardson-Dushman equation and is given by[10]

$$J = AT^2 e^{-eW_f/kT} \quad \text{A m}^{-2}$$

where A is a constant and T is the absolute temperature.

Typical materials used for cathodes are tungsten, thoriated tungsten and the barium and strontium oxides. Table 3 gives the value of the work function W_f and operating temperature T. Thermionic emitters are used extensively in electronic valves such as thermionic diodes, triodes and pentodes.

Photoelectric emission[20]

Electrons can be emitted when radiant energy falls on certain metallic surfaces. This is known as the photoelectric effect and energy is supplied according to Planck's law $W = hf$. For an electron to leave the surface with kinetic energy $\frac{1}{2}mv^2$, the relationship to be satisfied is given by the Einstein equation

$$hf \geqslant eW_f + \tfrac{1}{2}mv^2$$

where W_f is the work function of the material. Hence, the critical condition for photoelectrons to *just* leave the surface with zero velocity is given by

$$f_c = \frac{eW_f}{h}$$

or

$$\lambda_c = \frac{c}{f_c} = \frac{hc}{eW_f}$$

where λ_c is the critical wavelength of the incident radiation.

It is interesting to note that the wavelength of the incident radiation affects the velocity of emission of the photoelectrons only, and to increase the number of emitted electrons, the *intensity* of the radiation must be increased. Typical materials used as photocathodes in photomultiplier devices are antimony or strontium oxides. Other devices such as the photovoltaic cell (solar cell) use silicon, while the photoconductive cell uses cadmium sulphide.

Secondary emission

Electrons emitted by the photoelectric effect may be used to emit *secondary* electrons, thereby increasing the current through the device such as in the photomultiplier. This multiplication process is due to the kinetic energy of the primary photoelectrons which on hitting another photocathode are able to knock out one or more secondary electrons from the surface. The ratio of the number of secondary electrons to the number of primary electrons is called the secondary emission yield. Typically, for a primary electron energy of about 500 eV, this ratio is around four to ten for the alkali oxides and as high as twenty for alkali halides.

High-field emission

The presence of a high electric field near a metallic surface can pull out some of the surface electrons even at room temperature. This phenomenon is known as high-field emission and is due to the presence of a very thin barrier potential at the surface. According to classical theory, electrons would have to surmount this barrier to be emitted, but due to quantum theory, an electron can behave as a wave which may penetrate through the barrier so as to appear outside the surface. This effect is also known as *tunnelling*[20] and is found to occur even at low voltages in certain semiconductor devices such as the tunnel diode and laser diode (see Section 4.1). This is because the field gradients produced are of the order of 10^8 V m^{-1} across the semiconductor *p-n* junction.

Field emission is usually found to occur in certain gas discharge valves. It can also occur in a heated vacuum diode at a high anode potential and it is then known as the *Schottky effect*.[10]

Example 15

Discuss the phenomenon of space-charge limitation of current in a thermionic diode.

The space-charge-limited current density J in a diode is known to be of form AV^n where V is the voltage across the diode and the perveance A is a function of ε_0 (permittivity of free space), e and m (charge and mass of an electron) and d (anode-to-cathode spacing). Using fundamental equations, determine the value of n and the nature of the dependence of A on ε_0, e, m and d.

State briefly the limitations of the result so obtained.

Solution

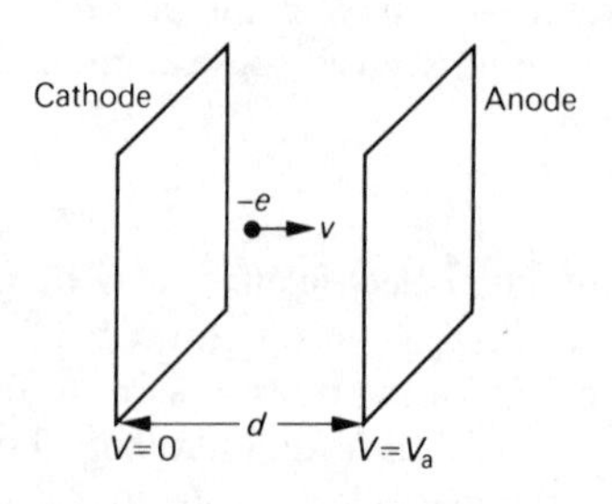

Fig. 60

For the plane geometry shown in Fig. 60, the current density J is given by

$$J = -\rho v$$

where ρ is the charge density, v is the electron velocity and the negative sign assumes that current flow is opposite to electron motion.

For an accelerating voltage V, the kinetic energy of the electrons of mass m and charge e is given by

$$\tfrac{1}{2} m v^2 = eV$$

or

$$v = \sqrt{\frac{2eV}{m}}$$

and

$$\rho = -\frac{J}{v} = -\frac{J}{\sqrt{2e/m}} V^{-1/2}$$

From Poisson's equation we obtain

$$\frac{d^2 V}{dx^2} = -\frac{\rho}{\varepsilon_0} = \frac{J}{\varepsilon_0 \sqrt{2e/m}} V^{-1/2}$$

where ε_0 is the permittivity of the vacuum.

Multiplying both sides by $2\,dV/dx$ and integrating yields

$$\left(\frac{dV}{dx}\right)^2 = \frac{4J}{\varepsilon_0 \sqrt{2e/m}} V^{1/2}$$

or

$$\frac{dV}{dx} = \left[\frac{4J}{\varepsilon_0 \sqrt{2e/m}}\right]^{1/2} V^{1/4}$$

and

$$\frac{V^{3/4}}{x} = \tfrac{3}{4}\left[\frac{4J}{\varepsilon_0 \sqrt{2e/m}}\right]^{1/2}$$

as the constants of integration are zero if $V = 0$ at $x = 0$.

By squaring both sides and rearranging terms yields

$$J = \frac{4\varepsilon_0}{9} \sqrt{\frac{2e}{m}} \frac{V^{3/2}}{x^2}$$

and if $V = V_a$ at $x = d$ we obtain

$$J = \frac{4\varepsilon_0}{9d^2} \sqrt{\frac{2e}{m}} V_a^{3/2} = A V_a^n \quad \mathrm{A\,m^{-2}}$$

where

$$A = \frac{4\varepsilon_0}{9d^2} \sqrt{\frac{2e}{m}}$$

and

$$n = 3/2$$

Hence, the constant A reveals the relationship between ε_0, m, e and d explicitly.

Limitations

1. The result obtained is for a planar geometry while a cylindrical geometry is used more often in practice. However, this only alters the value of the constant A.
2. The field at the cathode is not zero but attains this value a short distance from it.
3. The electron velocity is not zero at the cathode but has a small initial value.

In spite of these limitations, the general form of the result, which is known as the *Child-Langmuir law*,[10] applies quite well in practice.

Example 16

Discuss briefly the various processes responsible for emission of electrons from surfaces.

A photocathode with a work function of 1·6 V is illuminated at 10·0 lux (lumen/m^2) by monochromatic radiation of 500 nm wavelength. The cathode has a thermionic emission constant of $2{\cdot}5 \times 10^4$ A/(m^{2}K^2). Calculate approximate values for

(a) the threshold wavelength of the cathode;
(b) the photoelectric emission current density assuming a quantum yield of unity;
(c) the thermionic-emission current density assuming the cathode temperature to be 300 K.
(Note: 680 lumens = 1 Watt) (C.E.I.)

Solution

The important processes are known as thermionic emission, photoelectric emission, secondary emission and high-field emission. Further details are given in Section 5.3.

Problem

(a) The threshold wavelength is given by

$$\lambda_0 = hc/e\phi$$

or

$$\lambda_0 = \frac{6{\cdot}625 \times 10^{-34} \times 3 \times 10^8}{1{\cdot}6 \times 10^{-19} \times 1{\cdot}6} = 776\,\text{nm}$$

(b) 10 lux = 10/680 = 0·0147 W m^{-2}

$$1 \text{ quantum} = hc/\lambda = \frac{6{\cdot}625 \times 10^{-34} \times 3 \times 10^8}{500 \times 10^{-9}} = 3{\cdot}975 \times 10^{-19}\ \text{joules}$$

$$\text{Number of electrons/s/m}^2 = \frac{0{\cdot}0147}{3{\cdot}975 \times 10^{-19}} = 3{\cdot}698 \times 10^{16}$$

with $$J = 3{\cdot}698 \times 10^{16} \times 1{\cdot}6 \times 10^{-19} = 5{\cdot}92 \times 10^{-3}\ \text{A m}^{-2}$$

(c) $$J = AT^2\, e^{-q\phi/kT}$$

$$\frac{q\phi}{kT} = \frac{1{\cdot}6 \times 10^{-19} \times 1{\cdot}6}{1{\cdot}38 \times 10^{-23} \times 300} = 61{\cdot}84$$

$$e^{-61{\cdot}84} = 10^{-26{\cdot}89}$$

and $$J = 2{\cdot}5 \times 10^4 \times 300 \times 300 \times 10^{-26{\cdot}89}$$

or $$J \simeq 2{\cdot}9 \times 10^{-18}\ \text{A m}^{-2}$$

6

Vacuum devices

Electronic devices whose proper operation depends upon a good vacuum are generally known as vacuum devices, in contrast with certain other devices which depend largely on the presence of a gas and are usually called gas-filled devices. Vacuum devices may be broadly grouped as thermionic devices, display devices, photon devices and microwave devices.

6.1 Thermionic devices

The most commonly used thermionic devices have been the diode, triode and pentode. They have been largely superseded for low power applications by semiconductor devices but are still used in many television receivers and in high power applications such as broadcast transmitters.

Vacuum diode

The vacuum diode consists of an evacuated region usually within a glass envelope, which contains two electrodes known as the anode and cathode. It is shown symbolically in Fig. 61(a).

The two electrodes are usually cylindrical in shape with the cathode in the centre and the anode surrounding it. The cathode is heated indirectly by a

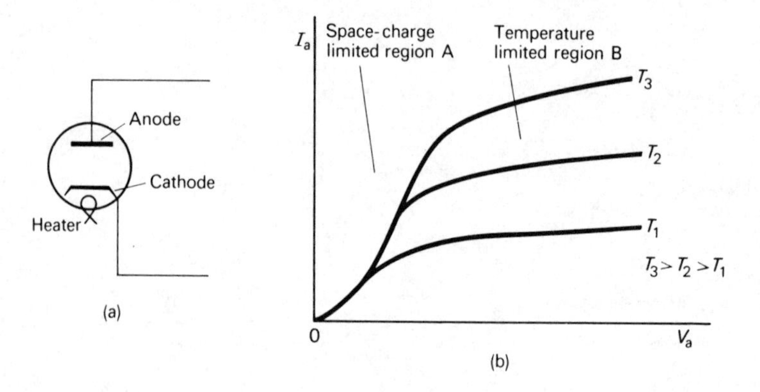

Fig. 61

filament wire and the emitted electrons are attracted by the positive potential applied to the anode. Diodes are used for various applications ranging from very small diodes for the demodulation of radio waves to the very large devices for power rectification.

A typical diode has the non-linear characteristic shown in Fig. 61(b) which shows the variation of anode current I_a with anode voltage V_a, for various cathode temperatures. They are called the anode characteristics of the diode and can be distinguished by a space-charge limited region A and a temperature-limited region B.

In the space-charge limited region A, the diode current is limited by the presence of a space-charge cloud of electrons near the cathode, which being negatively charged tends to repel some of the emitted electrons and so reduces the maximum current attainable. The I_a/V_a characteristic is given by the Child-Langmuir three-halves power law[10] which is of form

$$I_a = AV_a^{3/2} \quad \text{amps}$$

where A is a constant depending on the geometry and material of the electrodes and on their distance apart. In practice, this is the region most commonly used for diode applications.

In the temperature-limited region B, the diode current is limited by the temperature of the cathode, and the characteristic tends to flatten at higher anode voltages. Hence, a larger current can only be obtained by increasing the cathode temperature as shown in Fig. 61(b). This part of the characteristic is employed in a special device known as the noise diode, which is used as a *white-noise* source. Random fluctuations of the diode current are most pronounced in this region and generate shot-noise, especially if a directly heated tungsten cathode is used. This device finds applications for noise measurements in communication systems.

Example 17

Sketch the V/I characteristic for a thermionic vacuum diode and explain briefly the physical phenomena which occur in each of the regions of the characteristic.

A diode has an anode dissipation of 2 W when passing a steady current of 40 mA. Calculate the repetition frequency for (a) a rectangular and (b) a triangular current pulse of duration 1 μs and amplitude 10 A when each produce a mean anode dissipation of 5 W. Space-charge limited operation applies in each condition. (C.E.I.)

Solution

The answer to the first part of the question is given in Section 6.1.

Problem

$$\text{Anode dissipation} = VI$$

or

$$V = 2/(40 \times 10^{-3}) = 50 \text{ V}$$

(a) For a rectangular waveform we have

$$I_{\text{mean}} = \text{Peak current} \times \text{duty ratio}$$

or

$$I_{\text{mean}} = (10 \times 10^{-6})/T = 10^{-5}/T$$

where $T = 1/f_r$ and f_r is the repetition frequency.

Hence

$$VI_{\text{mean}} = 5$$

or

$$(50 \times 10^{-5})/T = 5$$

with

$$T = (50 \times 10^{-5})/5 = 10^{-4}$$

or

$$f_r = 10\,\text{kHz}$$

(b) For a triangular waveform we have

$$I_{\text{mean}} = [(1/2) \times \text{base} \times \text{height}]/T$$

or

$$I_{\text{mean}} = [(1/2) \times 10^{-6} \times 10]/T = (5 \times 10^{-6})/T$$

where $T = 1/f_r$ and f_r is the repetition frequency.

Hence

$$VI_{\text{mean}} = 5$$

or

$$(50 \times 5 \times 10^{-6})/T = 5$$

with

$$T = 5 \times 10^{-5}$$

or

$$f_r = 20\,\text{kHz}$$

Triode

The current flowing through a diode can be controlled by the addition of a third electrode between the anode and cathode, which is called the control *grid*. A symbolic arrangement is shown in Fig. 62(a) where the grid is shown dotted to represent a wire mesh which allows current to flow through it and also exercises control over it. The control is obtained by applying a suitable d.c. or a.c. voltage

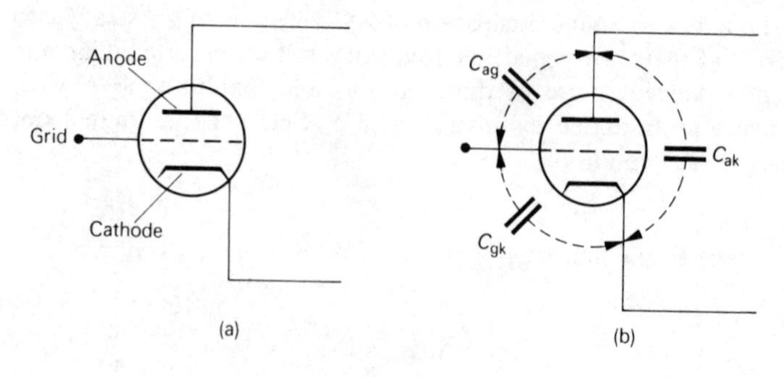

Fig. 62

to the grid. Such devices are used for various applications such as amplification, oscillation, mixing etc.

The two sets of characteristics of the triode are the *anode* characteristics relating anode current I_a and anode voltage V_a, and the *mutual* characteristics relating anode current I_a and grid voltage v_g. They are illustrated in Fig. 63 for a typical triode.

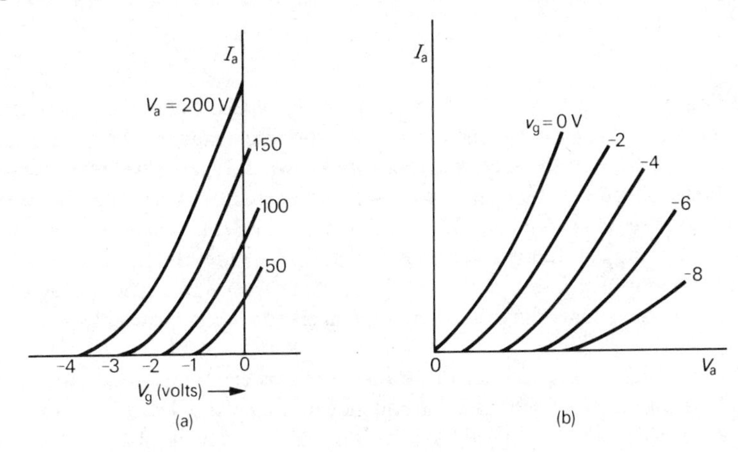

Fig. 63

The main parameters of the triode are its mutual conductance g_m and anode resistance r_a. The former is obtainable from the mutual characteristics and is expressed as the incremental change in I_a for an incremental change in grid voltage v_g, at a fixed anode voltage V_a. Hence

$$g_m = \left(\frac{\partial I_a}{\partial v_g}\right)_{V_a \text{ constant}}$$

where partial derivatives are used as I_a is a function of v_g and V_a. Graphically, g_m is the *slope* of the mutual characteristic for a given value of V_a as shown in Fig. 63(a).

The anode resistance or a.c. resistance r_a is obtainable from the anode characteristics and is expressed as the incremental change in V_a for an incremental change in I_a, at a fixed grid voltage v_g. Hence

$$r_a = \left(\frac{\partial V_a}{\partial I_a}\right)_{v_g \text{ constant}}$$

where partial derivatives are used also as I_a is a function of V_a and v_g. It is obtained graphically as the *slope* of the anode characteristic for a fixed grid voltage v_g as shown in Fig. 63(b). Typical values of g_m and r_a are given in Table 4 and also the amplification factor $\mu = g_m r_a$.

Table 4

Parameter	Triode	Pentode
g_m	3 mS	5 mS
r_a	10 kΩ	1 MΩ
μ	30	5×10^3

Triodes may be used at low or high power levels. Their main limitation occurs at high frequencies due to interelectrode capacitances and transit time effects. The interelectrode capacitances are shown in Fig. 62(b) where the most troublesome one is C_{ag}, the anode-grid capacitance which produces degenerative feedback from anode to grid. Due to the *Miller effect*, a capacitance of $(\mu + 1)C_{ag}$ appears across the input, and at high frequencies the input signal is severely damped, leading to a loss of gain. Moreover, the output capacitance C_{ak} also causes a loss of gain at high frequencies and limits the usable frequency range.

At still higher frequencies, the transit time of electrons from cathode to anode determines the maximum frequency at which the device may be used. Normally, the grid current and grid voltage are 90° out of phase, but at high frequencies a component of current appears in phase with the grid voltage giving rise to a conductance $G = ag_m T^2 \omega^2$ where a is a constant, g_m is the mutual conductance of the triode, T is the period of the input signal and ω is its angular frequency. Hence, the conductance term heavily damps the input signal at very high frequencies. The effect is only overcome by using special disc-seal triodes or by making proper use of the transit time effect as in the klystron or magnetron.

Pentode

The anode-grid capacitance C_{ag} can be considerably reduced by placing another electrode between the anode and grid, which is held at a suitable d.c. potential relative to the cathode and at *zero* a.c. potential by decoupling it to earth with a capacitor. It is known as the *screen* grid and this structure constitutes a *tetrode.* In a pentode, an additional grid called the *suppressor* grid is placed between the anode and screen. It is usually held at cathode potential to reflect back secondary electrons emitted by the anode. The construction is shown schematically in Fig. 64.

Like the triode, the pentode has two sets of characteristics usually known as the anode characteristics and the mutual characteristics, both of which are shown in Fig. 65. From these characteristics, the two important parameters of g_m and r_a are determined as for the triode. The value of g_m is usually higher than that for a triode, and r_a is considerably higher. Typical values are given in

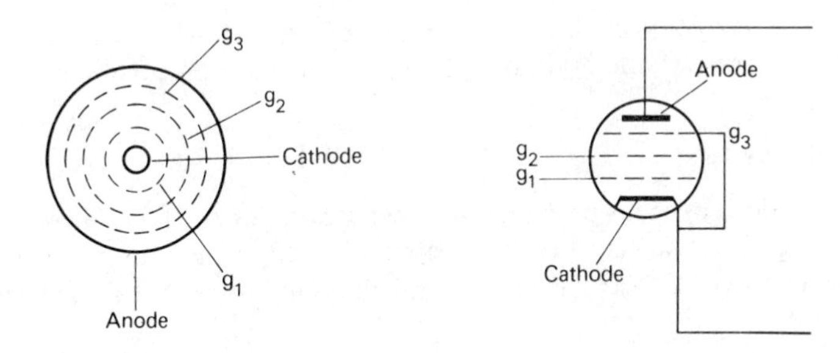

Fig. 64

Table 4. The expressions for g_m and r_a are given respectively by

$$g_m = \left(\frac{\partial I_a}{\partial v_g}\right)_{V_a \text{ constant}}$$

$$r_a = \left(\frac{\partial V_a}{\partial I_a}\right)_{v_g \text{ constant}}$$

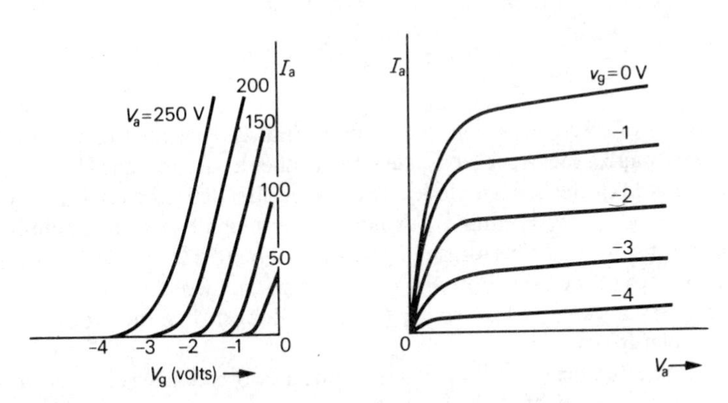

Fig. 65

Small pentodes are used extensively in voltage amplifiers, and larger pentodes are employed for power amplification because they produce low distortion. However, their high frequency performance is somewhat inferior to that of certain triodes and they also produce more noise, due to an additional effect known as *partition* noise.

6.2 Display devices

In electronic systems, information is often displayed visually as signal waveforms, television pictures or alphanumeric data. The most commonly used devices are the cathode ray tube, television tube and numerical indicators.

Cathode ray tube

The cathode ray tube is probably the most commonly used display device for examining signal waveforms. It consists essentially of an electron source, a focussing system, a deflecting system and a fluorescent screen. This is illustrated in Fig. 66.

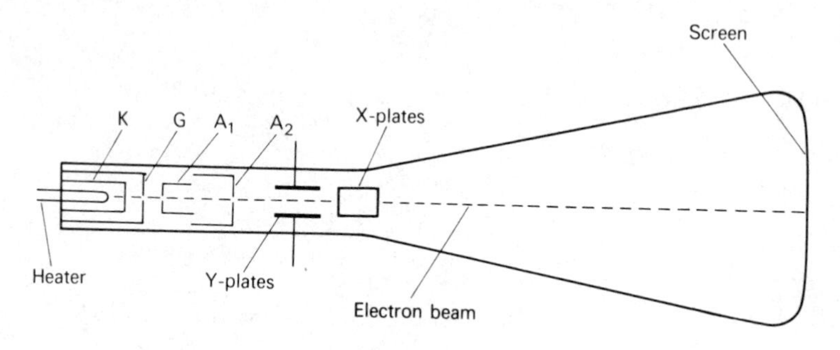

Fig. 66

A beam of electrons is emitted from an indirectly heated cathode K and passes through a focussing lens system to produce a narrow pencil beam, which is accelerated by the final anode A_2. The beam is first deflected vertically by a set of horizontal plates Y_1 and Y_2, which are kept small so as to minimise the transit time of electrons through them, as this affects the high frequency performance of the tube. For horizontal deflection, these are followed by a set of vertical plates X_1 and X_2 which are usually longer than the Y plates to give greater sensitivity.

The deflected beam of electrons then hits a fluorescent screen placed some distance away and the kinetic energy of the electrons causes the screen phosphor to glow at the point of impact. A short persistence blue-green phosphor is used for most applications and occasionally post deflection acceleration may be employed to increase the brightness. The signal waveform to be examined is applied to the Y-plates and a time-base voltage to the X-plates. The latter is usually a linear sawtooth waveform whose speed can be varied in steps from a few nanoseconds to several microseconds for a full scan. The screen is illuminated during the forward scan but on fly-back the spot is blanked out so as not to reveal the return trace.

Instruments incorporating such a device are called oscilloscopes which also provide the necessary electronic circuits for operating the cathode ray tube, together with other facilities for obtaining repetitive or single-scan operation and also amplification of the Y-plate signal. In addition, voltage measurements can be performed on the Y-plate signal using a calibrated graticule on the front face of the instrument, while time measurements can be made on the horizontal scan.

A form of cathode ray tube which is capable of storing information for a limited period of time is the *storage* tube. The information is in the usual form of an electrical signal which is stored as charge on a dielectric surface. By using two electron guns, the information can be directly displayed on the fluorescent screen or subsequently erased when required.

Colour television tube

An alternative to electrostatic deflection is the use of electromagnetic coils for deflecting the electron beam, and cathode ray tubes used in television and radar employ this technique. For colour television, the shadow-mask tube[21] and the trinitron tube[22] are largely employed. The shadow-mask tube consists essentially of three electron guns, a perforated shadow-mask screen and a phosphor dot fluorescent screen. It employs electrostatic focussing and electromagnetic deflection as illustrated in Fig. 67.

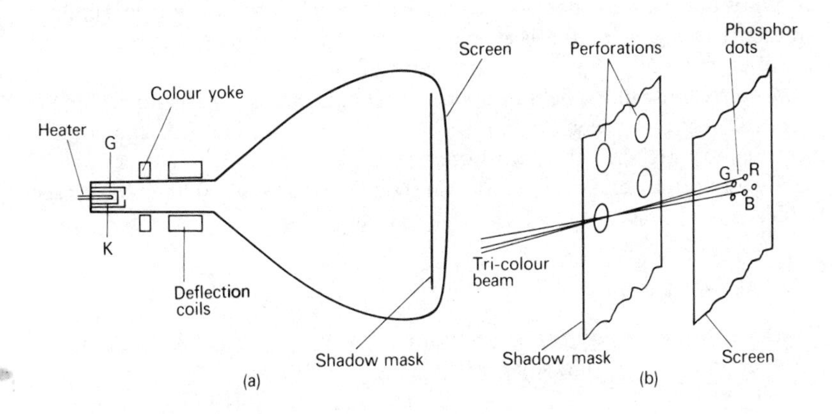

Fig. 67

The group of three electron beams, one for each colour red, green and blue, is directed towards the centre of the shadow-mask. The latter is a metal screen consisting of an array of perforations which are suitably spaced so that each electron beam can illuminate only its appropriate coloured phosphor dot on the fluorescent screen. The phosphor dots, which are in groups of the three colours red, green and blue, are arranged in the form of a triangular pattern as illustrated in Fig. 67(b). All the three beams can be directed to any part of the fluorescent screen by means of two pairs of deflecting coils or by the use of a single, large, toroidal winding.

In the trinitron tube, there is only one electron gun but it has three separate cathodes to produce three electron beams, one for each of the colours red, green and blue. The three beams are arranged to cross one another on the central axis, and the two outer beams diverge and are then electrostatically focussed to the same central point on the fluorescent screen as that of the centre beam. The three beams are also directed through a grill consisting of vertical slits suitably spaced so that each beam illuminates a particular coloured stripe on the fluorescent screen. The inner surface of the fluorescent screen is coated continuously with fine, vertical stripes of red, green and blue.

Light-emitting diode[23]

A *p-n* junction diode when forward biassed can be made to emit visible light, and is known as a light-emitting diode or LED, which is used extensively in pocket calculators. The radiation has a broad spectrum and is spontaneous and non-coherent. It is due to the recombination of electrons and holes which occur when conduction band electrons are captured by valence band holes.

By using suitable materials such as GaAs or GaP, the emission of red, green or yellow light can occur at low current densities. The phenomenon is therefore different from that of the laser diode, which requires much higher current densities and emits stimulated, coherent radiation.

More recently, in the field of optical communications, LED sources for the infra-red region are being employed. These are GaAs high-radiance diodes designed to match into the silica-fibre, optical cable. At present, power levels of nearly a milliwatt and modulation bandwidths of about 500 MHz have been achieved.

6.3 Photon devices

Three important devices using the photoelectric effect are the photomultiplier, photoconductive cell and solar cell.

Photomultiplier

When light falls on a photocathode, the emitted electrons may be only few in number. To increase the number several times, each primary electron is made to emit several secondary electrons by impact on another electrode at a higher potential. This is the principle of the photomultiplier shown in Fig. 68.

In Fig. 68, K is a photocathode coated with a material such as Cs_3Sb. Photons falling on K emit one or more electrons which are accelerated in turn to a series of shaped *dynodes* A_1, A_2 etc., held at successively higher potentials. Each dynode is coated with a low work-function material and emits as many as

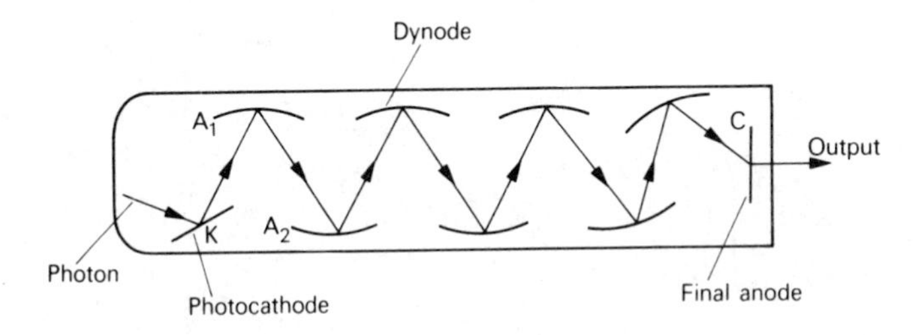

Fig. 68

ten secondary electrons for every primary electron. The dynodes are suitably positioned, and typically ten may be employed, giving an overall amplification factor of 10^{10} electrons per primary electron. The emitted electrons are collected at the final anode C to produce a large photoelectric current through an external circuit. An important application of such a device is the scintillation counter which is used for recording radioactive particles.

Photoconductive cell

The photoconductive cell is a device for detecting the amount of light radiation present. Typical applications are in photography and as infra-red detectors. For photographic purposes, the photoconductive cell consists of a pair of electrodes embedded in a CdS surface and connected externally to a small voltage. The spectral response of CdS corresponds to that of the normal eye and so it records the amount of daylight falling on it. It can be calibrated to indicate the amount of exposure required for a certain film.

As an infra-red detector,CdS has too slow a response and so a commonly used material is PbS, which is very sensitive to infra-red radiation and can be usefully employed in military applications such as missiles for destroying jet aircraft. A typical CdS device is shown in Fig. 69(a).

A semiconductor *p-n* junction when exposed to light can also be made to change its conductivity. Such a device is known as a photodiode and consists of a *p-n* junction which is reverse biassed. Photodetection occurs in the depletion region where the high electric field separates the electron-hole pairs which are excited on the absorption of radiation.

For sufficiently high fields, secondary electron-hole pairs are produced by impact ionisation, resulting in internal gain, as is the case in the avalanche photodiode. Such a silicon photodiode is best suited to use as a detector in an optical fibre system, employing wavelengths up to 1 μm.

Solar cell[24, 25]

The direct conversion of solar energy into electrical energy has been of

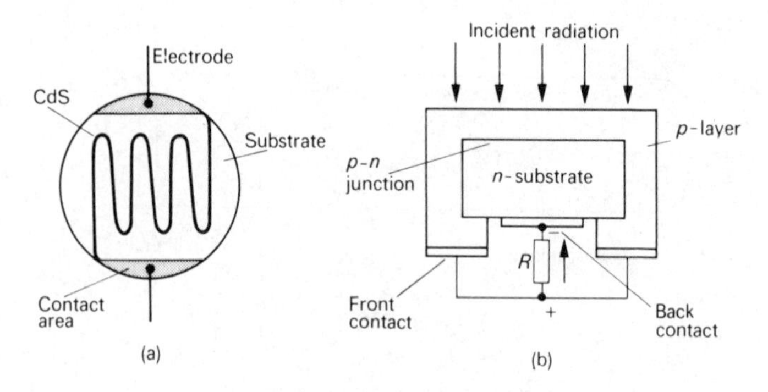

Fig. 69

considerable interest, especially in satellite communications for recharging conventional batteries. Hence, in recent years, much research and development has been done to produce a cheaper and more efficient device generally known as a 'solar cell'.

A typical solar cell is a photovoltaic device, since it generates a voltage when energised by sunlight. If a load resistor is connected in the external circuit, a current will flow through it as illustrated in Fig. 69(b). The solar cell is essentially a silicon *p-n* junction diode suitably doped and provided with a large surface area for collecting solar energy. The planar junction is thin and the exposed surface is given a suitably thin coating to reduce reflected light. Contacts to the *n*-region are made by 'metal fingers', while the *p*-material is mounted directly on a metallic surface.

For higher outputs, several solar cells are interconnected in a suitable way and mounted on the side of the satellite body or attached to long booms like that of a windmill. The efficiency of a typical solar cell is about 15% and it can generate an open-circuit voltage of about 0·5 V or a short-circuit current of about 100 μA in bright sunlight.

Example 18

Contrast and compare the characteristics of a thermionic pentode with those of a metal-oxide field-effect transistor. Sketch characteristics for each and give typical values for their parameters.

Draw a circuit for either a cathode-follower or a source-follower and derive relationships for its gain and output impedance. Suggest an application for a circuit of this type. (C.E.I.)

Solution

Contrast

1. The pentode operates with much higher voltages than the MOSFET.

2. The input impedance of an MOSFET is much higher than that of a pentode.
3. The pinch-off effect in an MOSFET is absent in a pentode.
4. The MOSFET has fewer electrodes than the pentode.

Comparison

1. The output characteristics of both devices are very similar, i.e. output current is fairly constant over a range of supply voltages.
2. The output current of both devices can be decreased by a negative input voltage (if a depletion mode MOSFET is used).
3. Both devices use similar parameters such as g_m, r_a (pentode) or g_m, r_d (MOSFET).
4. The output impedance of both devices are fairly high.

Typical characteristics

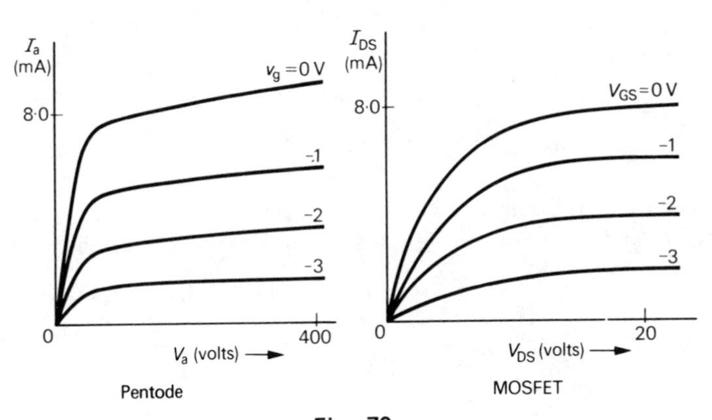

Fig. 70

Typical parameters

Pentode	*MOSFET*
$g_m = 5{\cdot}\text{mS}$	$g_m = 2{\cdot}5\ \text{mS}$
$r_a > 100\ \text{k}\Omega$	$r_d \simeq 100\ \text{k}\Omega$
$R_{in} = 10^6\ \Omega$	$R_{in} \simeq 10^{14}\ \Omega$

Fig. 71(a) shows a source-follower employing an *n*-channel (enhancement mode) MOSFET, and its equivalent circuit is shown in Fig. 71(b). The input resistors R_1 and R_2 set the bias for the gate and the output voltage appears across the source resistor R_s. For the equivalent circuit we obtain

$$v_i = v_{gs} + v_o$$

or

$$v_{gs} = v_i - v_o$$

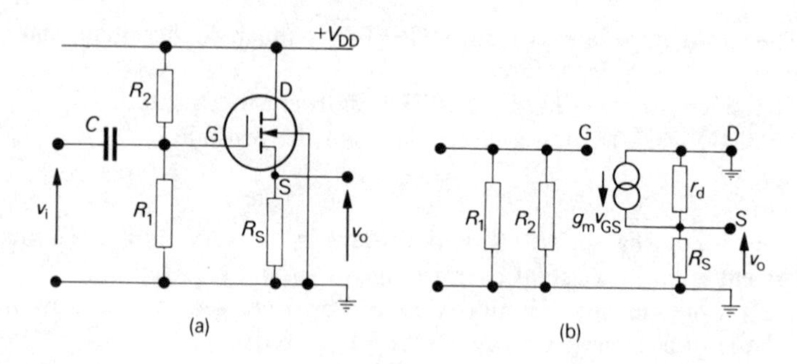

Fig. 71

Since r_d is generally large, the output voltage v_o is given by

$$v_o \simeq g_m v_{gs} R_s \simeq g_m R_s (v_i - v_o)$$

or

$$\frac{v_o}{v_i} \simeq \frac{g_m R_s}{1 + g_m R_s}$$

with

$$\text{Gain} = \frac{v_o}{v_i} = \frac{g_m R_s}{1 + g_m R_s}$$

To determine the output resistance R_o, a voltage v_s is applied at the output resistor R_s and the input voltage v_i is shorted. The output current is given by

$$i_o = v_s/r_d + v_s/R_s + g_m v_s = v_s(1/r_d + 1/R_s + g_m)$$

or

$$R_o = \frac{v_s}{i_o} = \frac{1}{(1/r_d + 1/R_s + g_m)}$$

6.4 Microwave devices[26]

The generation or amplification of energy at microwave frequencies of 1 GHz or more, requires special techniques to overcome transit time effects. The most commonly used devices which can achieve this are the klystron, magnetron and travelling wave amplifier.

Two-cavity klystron

The two basic forms of klystron used are the two-cavity klystron and the reflex klystron. In the two-cavity klystron shown in Fig. 72(a), an electron source emits a beam of electrons which are accelerated across a gap coupled to one resonant cavity called the *buncher* and are then allowed to drift through a second resonant cavity called the *catcher*, before finally being collected by the anode. The device may be operated as an amplifier or oscillator.

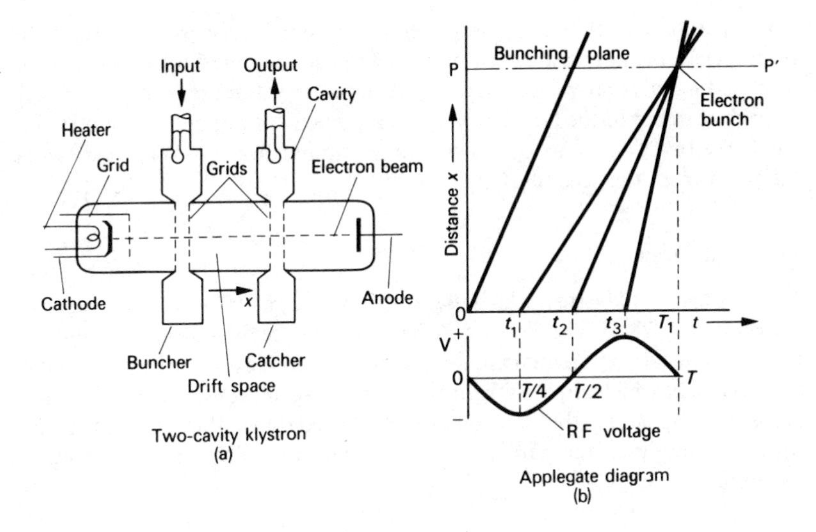

Fig. 72

The operation of the klystron depends on the principle of velocity modulation, whereby the uniform beam of electrons leaving the cathode is subsequently made to form 'bunches' of electrons, which is known as *bunching*. To understand how this bunching action arises, consider the Applegate diagram shown in Fig. 72(b), in which electrons are shown arriving at the first gap at instants t_1, t_2, t_3 of the RF voltage, which is assumed initially to exist at the first gap due to random noise fluctuations which are present in the stream of moving charge.

At time t_2, electrons move through the first gap when the RF voltage is zero, so they are passing through with a uniform velocity represented by the *slope* of the line at time t_2. At time t_1, the electron entering the first gap is decelerated by the negative RF voltage, while at time t_3, the electron entering the first gap is accelerated by the positive RF voltage. The corresponding velocities are given by the slopes of the lines drawn at instants t_2, t_1, t_3 respectively. Hence, it is observed that the electrons tend to arrive in a bunch some distance beyond the first gap. This grouping of electrons at the plane PP′ shows that the electrons are made to bunch at various times such as T_1.

The bunches of electronic charge are then allowed to drift through a field-free space known as the 'drift space', during which time faster moving electrons catch up with slower moving electrons emitted earlier in the RF cycle. Hence, the bunching action is increased in the drift space and subsequently the current bunches pass through the second gap. As this gap is coupled to another resonant cavity, the moving bunches induce a large voltage across it and are finally collected at the anode.

To operate the device as an amplifier, a magnetic loop is used to couple the input signal into the first cavity, and a similar loop is used for extracting the amplified signal from the second cavity. Alternatively, if some energy is fed back from the catcher to the buncher to sustain the RF voltage across the first gap, the device functions as an oscillator and continuous wave (CW) power can be extracted from the second cavity.

Reflex klystron

Another form of klystron which uses only a single cavity is the *reflex* klystron shown in Fig. 73(a). The electrons, after entering the drift space, are reflected back by a negative potential applied to the final anode and on returning, they form bunches which move back through the gap again. Hence, if the RF voltage across the gap is such that the bunches are slowed down, they are made to give up their energy to the gap thereby increasing its voltage and maintaining oscillations.

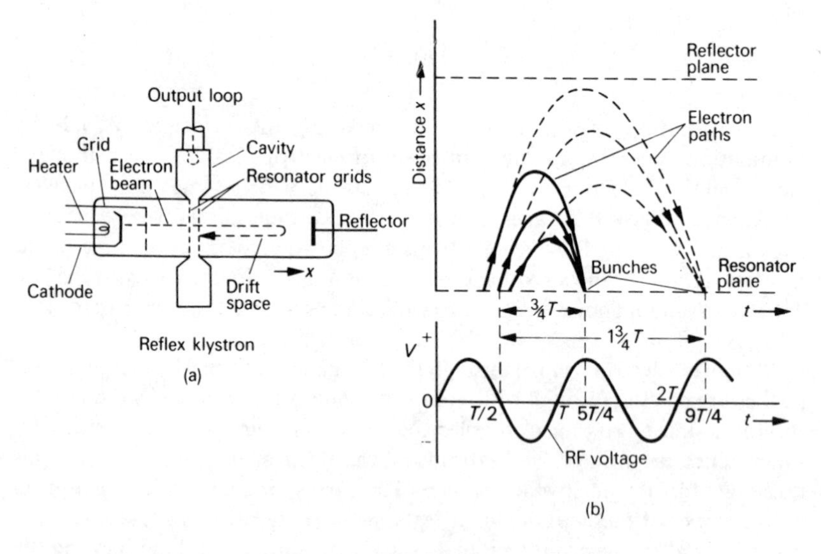

Fig. 73

The condition for oscillations to occur can be determined by considering the formation of the bunches as they pass back through the resonator gap. This is illustrated in Fig. 73(b) for the case of three electrons entering the gap for the first time and then being reflected back by the reflector voltage. During this time, the corresponding cycles of RF voltage are shown in Fig. 73(b) and are such that bunching occurs after time intervals of $(\frac{3}{4})T$, $(1\frac{3}{4})T$, etc. which correspond to its various *modes* of oscillation.

The lowest mode corresponding to the shortest bunching time and highest (negative) reflector voltage is the $\frac{3}{4}$ mode. The frequency and power output of the klystron vary as the reflector voltage is changed, and most of the power can be coupled out of the $1\frac{3}{4}$ mode as normally power cannot be coupled out of the $\frac{3}{4}$ mode due to the conductance of the system.

Klystron oscillators are used extensively as low power microwave sources with outputs from a few milliwatts to about one watt. For higher powers, multicavity klystron amplifiers are used delivering a few kilowatts (CW) or several megawatts (pulsed). Bandwidths are generally narrow and around 1–2 %, but recently wider bandwidths approaching 10 % have been attained for power applications in satellite ground stations or troposcatter systems.

Magnetron

A source of high power microwave energy is the multicavity magnetron shown in Fig. 74. It is basically a cylindrical diode with the cathode at the centre and an anode, consisting of several interconnected resonant cavities, placed around it. A steady axial magnetic field is applied in a direction parallel to the cathode and a d.c. or pulsed voltage is applied between anode and cathode.

Due to the combined action of the electric and magnetic fields, the electrons generally follow cycloidal paths to the anode and their d.c. energy is converted to RF energy at the cavities. Electrons which do not reach the anode spiral back and hit the cathode with considerable kinetic energy, which increases the temperature of its surface. Due to this back bombardment, the heater supply in high power magnetrons may be switched off during operation to avoid overheating.

The most normal mode of operation is the π-mode when alternate segments are in phase. If the electron motion is correct, it will reach the anode segments with little d.c. energy as it will have lost most of its energy to the RF field across the oscillating cavities. To ensure the π-mode operation, alternate segments are joined by metal straps, which is called *strapping*.

The exact build-up of oscillations is not clear but is believed to be associated with the instability of the space-charge cloud around the cathode, which rotates and breaks up into spoke-like arms reaching out to the anode block. This bunching of the space-charge cloud maintains the RF field across the cavities and power can be coupled out from a cavity by means of a magnetic loop.

The performance of a magnetron may be studied by means of a performance chart and a Rieke diagram which are shown in Fig. 75. The performance chart shows how the applied voltage and current vary, for various values of applied magnetic field at a particular load impedance, while the Rieke diagram shows how the power output varies with frequency of operation under various load conditions. The latter is essentially a Smith Chart with the voltage standing-wave ratio (VSWR) circles drawn in while the reactance circles are omitted. In this connection, the *pushing* figure of a magnetron is the change of frequency

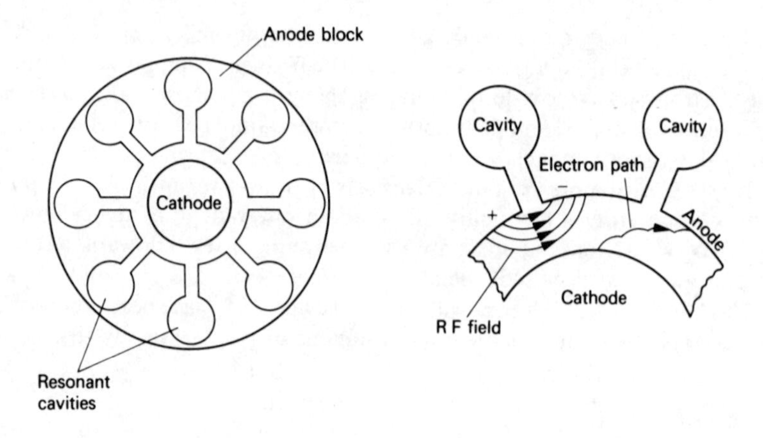

Fig. 74

for a defined anode voltage variation at a given load, which is obtained from the performance chart, and the *pulling* figure is the change of frequency for a given change in the standing-wave ratio at the output of the magnetron and is obtained from the Rieke diagram.

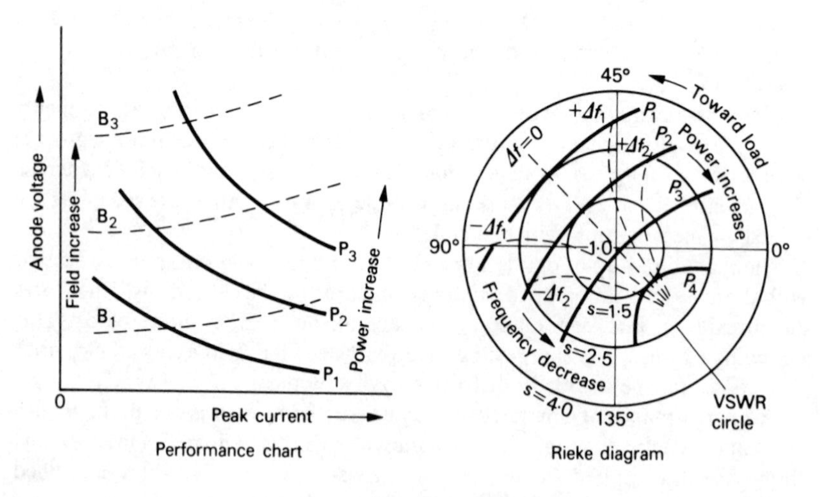

Fig. 75

A further development of the cavity magnetron is the coaxial magnetron which achieves good performance and stability. It uses a surrounding coaxial cavity in which the TE_{011}* coaxial mode is excited. The cavity provides a stabilising influence on the operating frequency and it is less affected by frequency pushing and pulling.

Magnetrons designed for continuous wave operation can deliver powers of about 1 kW at 1 GHz or a few watts at about 30 GHz. Pulsed magnetrons which are used extensively in radar are capable of providing peak powers of a few megawatts at about 1 GHz with typical efficiencies around 40%.

Travelling-wave amplifier[27, 28]

To overcome the limitation of the gain-bandwidth product associated with resonant circuits, the travelling-wave amplifier tube or TWT employs a helical slow-wave structure which has broadband properties and a beam of electrons which is velocity modulated. The beam of electrons from an electron source passes down the centre of the helix while the RF signal to be amplified is made to travel along the helix. In the arrangement shown in Fig. 76, the input signal is coupled into and out of the slow-wave structure by coaxial connectors and there is mutual interaction between the electron beam and the RF wave if the phase velocity of the beam is slightly higher than that of the RF wave.

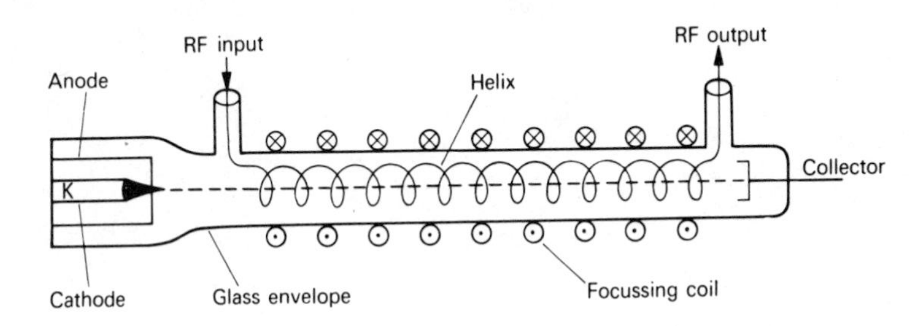

Fig. 76

In practice, the phase velocity of the wave is usually about one tenth the velocity of light and due to the continuous interaction, the input signal receives energy from the electron beam and an amplified signal with high gain can be obtained over a broad bandwidth. The mechanism of energy exchange is very similar to that of the klystron. The RF wave on the slow-wave structure causes electron bunching all along its length and if the bunches are slightly ahead of the RF field, they are retarded and give up their kinetic energy to the RF wave before being trapped at the final electrode or collector.

To maintain a strongly focussed beam along the centre of the helix, an external magnetic focussing field is required, which may be produced by a

* See F. R. Connor, *Wave Transmission*, p. 102, Edward Arnold (1972).

solenoidal winding or more usually by permanent magnets which employ periodically reversing fields to focus the beam.

The frequency behaviour of a TWT can be studied by means of an ω–β diagram where ω is the angular operating frequency and β is the phase shift per unit length of the slow-wave structure. The operating performance of the TWT is illustrated by its input/output characteristic which shows an initial linearity leading eventually to saturation effects. Hence, to obtain good linearity, the TWT is backed-off or operated about 2–3 dB below saturation. Furthermore, when several carriers are amplified simultaneously as in multicarrier operation, third order intermodulation products can occur which are only about 10 dB below the input carriers.

The TWT is used extensively in microwave links, satellite communications and in radar applications. Low power tubes provide outputs with 20–30 dB gain, octave bandwidths and efficiencies around 5–10 %. High power tubes have narrower bandwidths but yield efficiencies as high as 35 %. Operating frequencies generally cover the range 1 to 12 GHz.

Note

The light-emitting diode, photoconductive cell and solar cell are solid-state devices. They are included in this chapter for convenience.

Problems

1 Verify that the space-dependent part of the plane-wave function

$$\Psi = A\exp[\mathrm{j}(\omega t - k.r)]$$

is a solution of Schrödinger's time-independent wave equation for zero potential.

Use this solution to derive the density-of-states distribution function

$$N(E) = \frac{8\pi(2m^3E)^{1/2}}{h^3}$$

where E is the electron energy.

Discuss briefly the relevance of this result to the determination of the energy distribution of the carrier concentration in (a) a free-electron model of a metal (b) a semiconductor. (C.E.I.)

2 Discuss briefly the factors that control (a) the hole and electron densities (b) the hole and electron mobilities and (c) the Fermi level in a doped semiconductor under equilibrium conditions.

A specimen of silicon is doped with $1{\cdot}1 \times 10^{22}$ boron atoms/m^3 and 9×10^{21} phosphorus atoms/m^3. Calculate the electron and hole concentrations and the Fermi level. (Density of states in the valence band: $N_v = 1{\cdot}04 \times 10^{25}/\mathrm{m}^3$.) (C.E.I.)

3 Without mathematical detail show how Schrödinger's equation and Fermi-Dirac statistics may be used to develop a theoretical expression for the number of conduction electrons in a pure semiconductor.

A certain pure semiconductor has $1{\cdot}5 \times 10^{16}$ conduction electrons/m^3 at 300 K. Calculate the energy gap between valence band and conduction band, and the Fermi energy. The constant in the expression relating the number of conduction electrons per m^3 to temperature and energy levels is $4{\cdot}83 \times 10^{21}\ \mathrm{m}^{-3}\,(\mathrm{K})^{-3/2}$. (U.L.)

4 Describe what is meant by mobile and immobile charges in an impurity semiconductor and explain the formation of a depletion layer at a *p-n* junction.

An n-type semiconductor has 10^{14} donor atoms/cm^3 which are completely activated at 300 K. The electrons in the conduction band can be represented by $N_c = 2{\cdot}5 \times 10^{19}$ available levels/cm^3 at the bottom of the band E_c with a probability of occupation equal to $\exp[-(E_c - E_F)/kT]$.

Determine the position of the Fermi level. At 300 K, $kT = 0{\cdot}026$ eV.
(U.L.)

5 A sample of germanium has dimensions 1 cm long (x-direction), 2·0 mm wide (y-direction) and 0·2 mm thick (z-direction). A voltage of 1·4 V is applied across the ends of the sample and a current of 10 mA is observed in the positive x-direction. A Hall voltage of 10 mV is observed in the y-direction when there is a magnetic field of 0·1 T in the z-direction. Calculate (a) the Hall constant (b) the sign of the charge carriers (c) the magnitude of the carrier density and (d) the drift mobility of the carriers.
(U.L.)

6 Describe the Hall effect and define the terms *Hall constant* and *Hall angle*.
An n-type crystal of semiconductor is in the form of a bar 2 cm long, 1 mm deep and 2 mm wide. A 20 mV p.d. applied across the ends of the bar causes a current of 10 mA to flow along its length. A magnetic field of 0·1 T parallel to the 1 mm dimension of the bar produces a Hall voltage of 450 μV. Calculate the Hall constant, the carrier mobility and the carrier concentration.
(C.E.I.)

7 Indium is alloyed on a germanium crystal containing 10^{15} arsenic atoms per cm^3. After recooling, the alloyed region contains 10^{17} indium atoms per cm^3. Calculate the electrostatic potential difference across the p-n junction so formed when the sample is at room temperature (300 K). All impurities may be assumed to be ionised. The carrier product n_i^2 is $5{\cdot}8 \times 10^{26}/\text{cm}^6$ in germanium at 300 K.
(U.L.)

8 A p-n junction in germanium is formed by a p-type material having 10^{15} acceptor atoms/cm^3 and an n-type material with 10^{14} donor atoms/cm^3. Evaluate the difference in electrostatic potential across this junction in the equilibrium condition at 300 K, assuming that all the impurity atoms are fully ionised. The distribution function for the electron energies can be taken as $f(E) = \exp[-(E - E_F)/kT]$, the available energy levels for electrons in the conduction band can be represented by $N_c = 2{\cdot}5 \times 10^{19}$ levels/cm^3 situated at the bottom of the conduction band, $kT = 0{\cdot}026$ eV at 300 K, E_F is the Fermi level and the energy gap for germanium is 0·72 eV.
(U.L.)

9 Explain the basic breakdown mechanisms that can occur in a reverse-biassed p-n junction and discuss the use of them in a Zener diode used as a voltage reference.
A Zener diode is doped very much more heavily on its p-side so that $N_A \gg N_D$. Assuming that the reverse breakdown voltage is considerably greater than the contact potential, find an approximate expression for the breakdown voltage in terms of the breakdown strength of silicon and the diode parameters. If the diode has a donor doping density of $10^{23}/\text{m}^3$ and the breakdown strength of silicon may be assumed to be 6×10^7 V m^{-1}, estimate its reverse breakdown voltage using the same assumptions.
(C.E.I.)

10 Explain briefly why the capacitance of a varactor diode varies with the bias voltage.

A diode having a capacitance $C = (65/V^{1/3})$ pF, where V is in volts, is connected across the frequency-determining circuit of one of the RF oscillator stages in a beat-frequency oscillator. The b.f.o. gives an output frequency range of 0 to 20 kHz when the frequency of the RF oscillator is swept from 500 kHz to 520 kHz. The inductance in the tuned circuit is 600 μH. With a 3·5 V bias across the diode, determine (a) the capacitance that, connected in parallel with the diode and the inductor, will give an oscillation frequency of 500 kHz and (b) the bias to give an oscillation frequency of 520 kHz. (C.E.I.)

11 A bipolar transistor has a leakage current I_{CBO} of 1 μA when connected in a common-emitter circuit and a base current of 20 μA is required to produce a collector current of 2 mA. Calculate the low-frequency current gain of the transistor and the value of the leakage current I_{CEO}.

12 List the principal differences between a bipolar and a junction field-effect transistor.

The transfer characteristics of a JFET may be represented by the relationship

$$I_{DS} = I_{DSS}(1 - V_{GS}/V_p)^2$$

where I_{DS} is the saturation drain current, I_{DSS} is the value of I_{DS} when $V_{GS} = 0$, V_p is the pinch-off voltage and $V_{GS} \leq V_p$. Show that the mutual conductance varies as the square root of the drain current and plot the characteristic for a typical JFET.

An *n*-channel JFET has $V_p = -2$ V and $I_{DSS} = 1{\cdot}5$ mA. Calculate the value of the source bias resistance R_s for $I_D = 1$ mA if the gate is at earth potential. (C.E.I.)

13 The energy density of radiation in a laser cavity of length L may be assumed uniform throughout its volume. Output mirrors at the ends of the cavity have intensity coefficients for reflection and diffraction of R and D respectively. Obtain an approximate expression for the population inversion necessary for oscillation to start in terms of R, D, L, λ, τ and $\Delta\lambda$. Here λ and $\Delta\lambda$ are the laser transition wavelength and (essentially narrow) linewidth, respectively, and τ is the radiative lifetime of the upper state. Note: the Einstein coefficient for stimulated emission is $\lambda^3/8\pi h\tau$.

(C.E.I.)

14 Describe the mechanism of (a) photoelectric emission and (b) secondary emission of electrons from a surface.

A photomultiplier has a cathode with a work function of 1·5 eV and ten dynodes each with a secondary emission coefficient of six. If radiation is incident on the cathode, calculate (a) the maximum wavelength for which collector current will flow (b) the maximum initial electron velocity if the wavelength of the radiation is 0·6 μm and (c) the final collector current if the cathode current is 10^{-10} A. (U.L.)

15 The Child-Langmuir equation for the space-charge limited current in a planar diode of area A and electrode spacing d is

$$I = \frac{4}{9}\varepsilon_0 \left(\frac{2e}{m}\right)^{1/2} V_a^{3/2} \frac{A}{d^2}$$

where the other symbols have their usual meaning. It is found experimentally that for a planar triode, the anode current is given by the above expression with V_{eff} substituted for V_a and d_{eff} for d, where V_{eff} is equal to $V_g + V_a/\mu$. Show that μ is the amplification factor of the triode.

A planar triode, for which d_{eff} is 5×10^{-4} m, has a cathode current density of 200 A m^{-2} when the grid and the anode voltages are -3 V and $+200$ V respectively. Estimate the amplification factor of the triode. (U.L.)

16 Develop the theory of electric field deflection in a cathode-ray tube, showing how the sensitivity depends on the accelerator voltage and the deflector and tube geometry.

Calculate the maximum accelerator voltage that may be used in a tube having parallel deflector plates 4 mm apart and 18 mm long, the plate centres being 220 mm from the screen, if a peak-to-peak deflection of 75 mm is to be produced by a transistor amplifier having a maximum r.m.s. output of 15 V. (U.L.)

17 Electrons accelerated from rest through a potential difference of 3 kV are projected into an evacuated region in which exists a uniform transverse field. Determine the trajectory followed by the electrons if this field is (a) an electric field of field strength 8 V mm^{-1} and (b) a magnetic field having a flux density of 0·005 Wb m^{-2}. Derive all expressions from first principles. (U.L.)

18 Describe the main features of one type of klystron valve.

The voltage on the resonator of a certain klystron is 1 kV. The p.d. between the closely-spaced grids of the buncher is 200 V (max) at a frequency of 1 GHz. Estimate the shortest distance from the buncher at which the maximum bunching will occur. (U.L.)

19 Explain the principle and describe the constructional form of a travelling-wave tube for amplification at microwave frequencies.

The helix of a travelling-wave tube has 8 turns/cm, a diameter of 0·38 cm and an axial length of 23 cm. Estimate what final-anode voltage must be applied to obtain useful gain. Also calculate the average transit time of an electron through the helix. (U.L.)

Answers

2 $n = 1{\cdot}04 \times 10^{11}/\text{m}^3$, $p = 2{\cdot}0 \times 10^{21}/\text{m}^3$, $E_F = 0{\cdot}22\,\text{eV}$ above the valence band
3 1·1 eV, 0·54 eV
4 0·32 eV below E_F
5 $2 \times 10^{-3}\,\text{m}^3/\text{C}$, n-type, $3{\cdot}13 \times 10^{21}/\text{m}^3$, $0{\cdot}36\,\text{m}^2/\text{V-s}$
6 $0{\cdot}45 \times 10^{-3}\,\text{m}^3/\text{C}$, $2{\cdot}25\,\text{m}^2/\text{V-s}$, $n = 1{\cdot}39 \times 10^{22}/\text{m}^3$
7 0·31 V
8 0·134 V
9 23·3 V
10 126 pF, 10·17 V
11 95, 100 μA
12 368 Ω
14 0·83 μm, $4{\cdot}5 \times 10^5\,\text{m s}^{-1}$, 6·1 mA
15 18·7
16 280 V
17 $y = 0{\cdot}67\,x^2$ (parabolic), radius = 3·7 cm (circular)
18 5·5 cm
19 2·87 kV, 7·2 ns

References

1 BARDEEN, J. and BRATTAIN, W. H. *Physical Review*, **70**, 230, 1948.
2 SHOCKLEY, W., SPARKS, M. and TEAL, G. K. *p-n* junction transistors. *Physical Review* **83**, 151, 1951.
3 SHOCKLEY, W. *Proceedings Institute of Radio Engineers*, **40**, 1365, 1952.
4 WEIMER, P. K. *et al. Proceedings Institute of Electrical and Electronic Engineers*, **52**, 1479, 1964.
5 ALTMAN, L. *Microprocessors.* McGraw-Hill (1975).
6 KERWIN, L. *Atomic Physics.* Holt, Rinehart and Winston (1963).
7 JAMMER, MAX. *The Conceptual Development of Quantum Mechanics.* McGraw-Hill (1966).
8 SCHIFF, L. I. *Quantum Mechanics.* McGraw-Hill (1968).
9 SZE, S. M. *Physics of Semiconductor Devices.* John Wiley (1969).
10 VAN DER ZIEL, A. *Solid State Physical Electronics.* Prentice-Hall (1976).
11 STREETMAN, B. G. *Solid State Electronics.* Prentice-Hall (1972).
12 ZENER, C. *Proceedings Royal Society* (*London*), Series A, **145**, 523, 1934.
13 McKAY, K. G. Avalanche Breakdown in Silicon. *Physical Review*, **94**, 877, 1954.
14 ESAKI, L. *Physical Review*, **109**, 603, 1958.
15 MOLL, J. L. *Proceedings Institute of Radio Engineers*, **43**, 1807, 1955.
16 HOFSTEIN, S. R. and HEIMAR, F. P. The Silicon Insulated-Gate Transistor. *Proceedings Institute of Electrical and Electronic Engineers*, **51**, 1190, 1963.
17 SIEGMAN, A. E. *Microwave Solid-State Masers.* McGraw-Hill (1964).
18 *Reference Data for Radio Engineers*, Sixth Edition. Howard W. Sams & Co (1977).
19 PASZKOWSKI, B. *Electron Optics.* Iliffe Books Ltd (1968).
20 LEVINE, S. N. *Quantum Physics of Electronics.* Macmillan (1965).
21 LAW, H. B. A Three-Gun Shadow Mask Color Kinescope. *Proceedings Institute of Radio Engineers*, **39**, 1186, 1951.
22 YOSHIDA, S. and OHKOSHI, A. The Trinitron – A New Color Tube. *Institute of Electrical and Electronic Engineers, Transactions on Broadcast and TV Receivers*, **BTR-14**, 19, 1968.
23 MELCHIOR, H. *et al.* Photodetectors for Optical Communication Systems. *Proceedings Institute of Electrical and Electronic Engineers*, **58**, 1466, 1970.
24 CHAPIN, D. M. *et al. Journal Applied Physics*, **25**, 676, 1954.
25 ALTMAN, M. *Elements of Solid-State Energy Conversion.* Van Nostrand (1969).
26 GIACOLETTO, L. J. *Electronic Designers Handbook.* McGraw-Hill (1977).
27 PIERCE, J. R. *Travelling Wave Tubes.* Van Nostrand (1950).
28 HUTTER, R. G. Travelling Wave Tubes. *Advances In Electronics and Electron Physics*, **6**, 371, 1954, Academic Press, N.Y.
29 GUNN, J. B. Solid State Communications, **1**, 88, 1963.
30 LEE, T. P. *et al. Institute of Electrical and Electronic Engineers Transactions on Electron Devices*, **ED-15**, 741, 1968.
31 DE LOACH, B. C. and SCHARFETTER, D. L. Device Physics of Trapatt Oscillators.

Institute of Electrical and Electronic Engineers Transactions on Electron Devices, **ED-17**, 9, 1970.

32 FINK, D. G. *Electronic Engineers Handbook*. McGraw-Hill (1975).

33 BOYLE, W. S. and SMITH, G. E. Charge-Coupled Semiconductor Devices. *Bell System Technical Journal*, **49**, 587, 1970.

Appendices

Appendix A: The elements

Atomic Number	*Element*	*Atomic Number*	*Element*	*Atomic Number*	*Element*
1	H	36	Kr	71	Lu
2	He	37	Rb	72	Hf
3	Li	38	Sr	73	Ta
4	Be	39	Y	74	W
5	B	40	Zr	75	Re
6	C	41	Nb	76	Os
7	N	42	Mo	77	Ir
8	O	43	Tc	78	Pt
9	F	44	Ru	79	Au
10	Ne	45	Rh	80	Hg
11	Na	46	Pd	81	Tl
12	Mg	47	Ag	82	Pb
13	Al	48	Cd	83	Bi
14	Si	49	In	84	Po
15	P	50	Sn	85	At
16	S	51	Sb	86	Rn
17	Cl	52	Te	87	Fr
18	Ar	53	I	88	Ra
19	K	54	Xe	89	Ac
20	Ca	55	Cs	90	Th
21	Sc	56	Ba	91	Pa
22	Ti	57	La	92	U
23	V	58	Ce	93	Np
24	Cr	59	Pr	94	Pu
25	Mn	60	Nd	95	Am
26	Fe	61	Pm	96	Cm
27	Co	62	Sm	97	Bk
28	Ni	63	Eu	98	Cf
29	Cu	64	Gd	99	Es
30	Zn	65	Tb	100	Fm
31	Ga	66	Dy	101	Md
32	Ge	67	Ho	102	No
33	As	68	Er	103	Lr
34	Se	69	Tm	104	Rf
35	Br	70	Tb	105	Ha

Appendix B: The Schrödinger equation

The wave property of any particle of matter such as an electron is represented by a wave function $\psi(x, y, z) = \psi$ for brevity. The behaviour of the particle is ascertained by determining this function using certain postulates. According to these postulates, the variables x, y, z representing the particle position in three dimensions remains unchanged in any transformation. However, other dynamical variables such as the momentum or energy are replaced by *operators* which operate on the function ψ. Since the particle cannot be precisely determined, the total probability of doing so in all configuration space is *normalised* to unity such that

$$\int_{-\infty}^{\infty} \psi^* \psi \, d\tau = 1$$

where ψ^* is the conjugate of function ψ and $d\tau = dx\,dy\,dz$. The product $\psi^*\psi = |\Psi|^2$ is the square of the wave amplitude and is always real. It is important because it gives the probability of finding the particle at a certain position or in a certain energy state.

To obtain Schrödinger's equation we use the classical equation for the total energy of a particle in a system which is of form

$$\frac{1}{2m}(p_x^2 + p_y^2 + p_z^2) + V = W$$

where p_x, p_y, p_z are the components of momentum in three dimensions, V is the potential energy, W is the total energy and m is the mass of the particle.

The classical equation can be converted into the wave equation by using the operators $p_x = (\hbar/j)\partial/\partial x$, $p_y = (\hbar/j)\partial/\partial y$, $p_z = (\hbar/j)\partial/\partial z$ where $j = \sqrt{-1}$. Hence, we obtain

$$-\frac{\hbar^2}{2m}\left[\frac{\partial^2}{\partial x^2} + \frac{\partial^2}{\partial y^2} + \frac{\partial^2}{\partial z^2}\right] + V = W \qquad (\hbar = h/2\pi)$$

or

$$\frac{-h^2}{8\pi^2 m}\left[\frac{\partial^2}{\partial x^2} + \frac{\partial^2}{\partial y^2} + \frac{\partial^2}{\partial z^2}\right]\psi + V\psi = W\psi$$

after operating on the function ψ. Hence, using the Laplacian operator $\nabla^2 = \left(\frac{\partial^2}{\partial x^2} + \frac{\partial^2}{\partial y^2} + \frac{\partial^2}{\partial z^2}\right)$ and rearranging terms yields

$$\nabla^2\psi + \frac{8\pi^2 m}{h^2}(W - V)\psi = 0$$

which is Schrödinger's time-independent wave equation.

Comment

In some phenomena, the *time-dependent* Schrödinger wave equation must be used by replacing $\psi(x, y, z)$ with $\psi(x, y, z, t)$ and W with the operator $(-\hbar/\mathrm{j})\partial/\partial t$. Hence, we obtain the corresponding Schrödinger time-dependent wave equation as

$$\nabla^2\psi - \frac{8\pi^2 m V\psi}{h^2} + \mathrm{j}\frac{4\pi m}{h}\frac{\partial\psi}{\partial t} = 0$$

Appendix C: Fermi-Dirac distribution

The statistical distribution of particles such as electrons in a metal may be obtained quantum mechanically. The probability of an electron occupying a quantum energy level W_i is given by the number of ways of arranging N_i electrons in S_i states. Since the electrons are indistinguishable, the Pauli exclusion principle permits only combinations and not permutations. Hence, the total number of ways W is given by

$$W = \prod_i \frac{S_i!}{N_i!(S_i - N_i)!}$$

The above expression can be shown to be maximised when

$$N_i = \frac{S_i}{1 + \mathrm{e}^{(\alpha + \beta W_i)}}$$

and the probability $P(W) = N_i/S_i$ is given by

$$P(W) = \frac{1}{1 + \mathrm{e}^{(\alpha + \beta W)}}$$

where W represents any energy level within an interval $\mathrm{d}W$ and α, β are constants chosen such that

$$\alpha + \beta W_{\mathrm{F}} = 0$$

or

$$W_{\mathrm{F}} = -\alpha/\beta$$

Hence, W_{F} is defined as the *Fermi level* such that $P(W_{\mathrm{F}}) = 0{\cdot}5$, i.e. it is the energy level whose probability of being occupied is 0·5.

Assuming the electron gas behaves as an ideal gas, thermodynamical reasoning then yields the result $\beta = 1/kT$ where k is Boltzmann's constant and T is the absolute temperature of the electron 'gas'. Hence

$$P(W) = \frac{1}{1 + \mathrm{e}^{\beta(W - W_{\mathrm{F}})}}$$

or

$$P(W) = \frac{1}{1 + \mathrm{e}^{(W - W_{\mathrm{F}})/kT}}$$

Density distribution

To obtain the density distribution of electrons let $\mathrm{d}N$ be the number of electrons in the energy range $\mathrm{d}W$ and $\mathrm{d}S$ the corresponding number of quantum states. Hence

$$\mathrm{d}N = P(W)\,\mathrm{d}S$$

or

$$\frac{\mathrm{d}N}{\mathrm{d}W} = P(W)\frac{\mathrm{d}S}{\mathrm{d}W}$$

For electrons with two spin states, the number of quantum states S is given by a consideration of particles in a three-dimensional potential well. Hence, we obtain per unit volume

$$S = \frac{8\pi\,(2mW)^{3/2}}{3h^3}$$

and the number of states per unit energy is given by

$$\frac{\mathrm{d}S}{\mathrm{d}W} = \frac{4\pi(2m)^{3/2}}{h^3}W^{1/2}$$

and

$$\frac{\mathrm{d}N}{\mathrm{d}W} = P(W)\frac{\mathrm{d}S}{\mathrm{d}W} = \frac{4\pi(2m)^{3/2}W^{1/2}}{h^3[1+\mathrm{e}^{(W-W_F)/kT}]}$$

which is called, the density of electrons and is shown in Fig. 19(b).

To determine the number of electrons in the energy states below W_F when $T=0$, we note that $P(W)=1$ and we obtain

$$N = \int_0^{W_F}\frac{\mathrm{d}N}{\mathrm{d}W}\mathrm{d}W = \int_0^{W_F}\frac{\mathrm{d}S}{\mathrm{d}W}\mathrm{d}W$$

$$= \int_0^{W_F}\frac{4\pi(2m)^{3/2}W^{1/2}\mathrm{d}W}{h^3}$$

or

$$N = \frac{8\pi(2m)^{3/2}W_F^{3/2}}{3h^3} = \frac{8\pi(2mW_F)^{3/2}}{3h^3}$$

Electron and hole concentrations

For a semiconductor it is found that the density of states $(\mathrm{d}S/\mathrm{d}W)$ in the conduction band is proportional to $(W-W_c)^{1/2}$ where W_c is the lowest energy level in the conduction band and we obtain

$$\frac{\mathrm{d}S}{\mathrm{d}W} = A(W-W_c)^{1/2}$$

where A is a constant of proportionality. Hence, the density of electrons n in the conduction band is given by

$$n = \int_{W_c}^{\infty} \frac{dN}{dW} dW = \int_{W_c}^{\infty} P(W) \frac{dS}{dW} dW$$

or

$$n = A \int_{W_c}^{\infty} \frac{(W - W_c)}{(1 + e^{(W - W_F)/kT})} dW$$

Substituting $v = \dfrac{(W - W_c)}{kT}$ yields a standard integral and from tables we obtain the result

$$n = n_e e^{-(W_c - W_F)/kT}$$

where $n_e = 2\left[\dfrac{(2\pi m_e^* kT)}{h^2}\right]^{3/2}$ and m_e^* is the effective electron mass.[11] The expression for n_e may also be written in the form

$$n_e = AT^{3/2}$$

where $A = 2\left[\dfrac{2\pi m_e^* k}{h^2}\right]^{3/2}$ is a constant. Hence, we obtain

$$n = AT^{3/2} e^{-(W_c - W_F)/kT} = AT^{3/2} e^{-W_g/2kT}$$

where W_g is the width of the forbidden energy gap.

For the case of the hole concentration p, we replace $P(W)$ by $[1 - P(W)]$, and W_c by W_v, the highest level of the valence band. Since the valency levels can be considered as extending from $-\infty$ to W_v, it is found that the density of quantum states is given by

$$\frac{dS}{dW} = A(W_v - W)^{1/2}$$

where A is a constant of proportionality. Hence

$$p = \int_{-\infty}^{W_v} [1 - P(W)] \frac{dS}{dW} dW$$

or

$$p = \int_{-\infty}^{W_v} e^{(W - W_F)/kT} . A(W_v - W)^{1/2} dW$$

which on integration yields

$$p = p_e e^{-(W_F - W_v)/kT}$$

where $p_e = 2\left[\dfrac{2\pi m_h^* kT}{h^2}\right]^{3/2}$ and m_h^* is the effective hole mass.[11]

Appendix D: *p-n* junction diode

When a *p-n* junction is formed, the majority carriers diffuse across the junction and set up a barrier potential (see Section 3.5) which tends to prevent any further diffusion across the junction. However, some electrons in the *n*-region can surmount the barrier and move into the *p*-region, while in the *p*-region, electron-hole pairs are produced by thermal agitation and so some electrons in the *p*-region can also move into the *n*-region. A similar consideration shows that holes can also move in either direction across the junction.

In the equilibrium condition when no bias is applied, the charge carriers moving in either direction per unit area must be the same. Hence, we have

$$J_{en} = J_{ep}$$

$$J_{hn} = J_{hp}$$

where the subscripts *e*, *h* refer to electrons and holes respectively and *n*, *p* to the respective regions.

If a forward bias of V volts is now applied, then the majority carrier densities J_{en}, J_{hp} increase exponentially from their equilibrium value according to Maxwell-Boltzmann statistics, whereas the current densities J_{ep}, J_{hn} are hardly affected as they are determined by thermal effects and impurity concentrations only. Hence, the *net* electron and hole current densities at the junction are given by

$$J_e = J_{en} - J_{ep} = J_{ep}[e^{eV/kT} - 1]$$

$$J_h = J_{hp} - J_{hn} = J_{hn}[e^{eV/kT} - 1]$$

and the total current density at the junction is the sum of the electron and hole current densities and is given by

$$J = J_e + J_h$$

or

$$J = (J_{ep} + J_{hn})[e^{eV/kT} - 1]$$

When a large reverse bias is applied, the quantity in square brackets → − 1 and we obtain

$$J = -(J_{ep} + J_{hn}) = -J_s$$

where J_s is the reverse saturation current, which attains a maximum value determined by thermal effects and impurity concentrations only. Hence, in general

$$J = J_s[e^{eV/kT} - 1]$$

Alternatively, if the concentration of electrons in the *p*-region is n_p and no bias is applied, for equilibrium to be maintained, the rate of production of electrons must equal the rate at which electrons and holes recombine in the *p*-region. Hence, if the lifetime of an electron in the *p*-region is τ_e while diffusing over a diffusion length L_e, we must have

$$J_{ep} = \frac{en_p L_e}{\tau_e}$$

and similarly for holes with a concentration of p_n in the n-region, with a lifetime τ_h over a diffusion length L_p we must have

$$J_{hn} = \frac{ep_n L_h}{\tau_h}$$

Substituting $L_e = \sqrt{D_n \tau_e}$ and $L_h = \sqrt{D_p \tau_h}$ where D_n, D_p are the electron and hole diffusion coefficients respectively yields

$$J = e\left[\frac{n_p D_n}{L_e} + \frac{p_n D_p}{L_h}\right](\mathrm{e}^{eV/kT} - 1)$$

with $$J_s = e\left[\frac{n_p D_n}{L_e} + \frac{p_n D_p}{L_h}\right]$$

and if the junction area is A, the junction current is given by

$$I = JA = I_s[\mathrm{e}^{eV/kT} - 1]$$

where $$I_s = J_s A = eA\left[\frac{n_p D_n}{L_e} + \frac{p_n D_p}{L_h}\right]$$

Charge-storage effects

When forward bias is applied to a p-n junction diode, holes are injected into the n-region and are initially stored as *minority* charge-carriers near the junction. This charge-storage gives rise to a capacitative effect and is known as the *diffusion* capacitance.

If reverse bias is subsequently applied, the stored charge is withdrawn across the junction and a reverse current flows for a short time before it decays to the small leakage value at cut-off. Hence, this causes a small time-delay which limits the switching speed at which the diode can operate. For higher switching speeds, a Schottky-barrier diode may be used.

Appendix E: JFET and IGFET equations

JFET equation

For the purpose of obtaining the basic JFET equation, an n-channel device will be assumed in which the depletion regions are symmetrical and vary as shown in Fig. 77.

Since the gate regions are heavily doped we have $N_A \gg N_D$ and the depletion

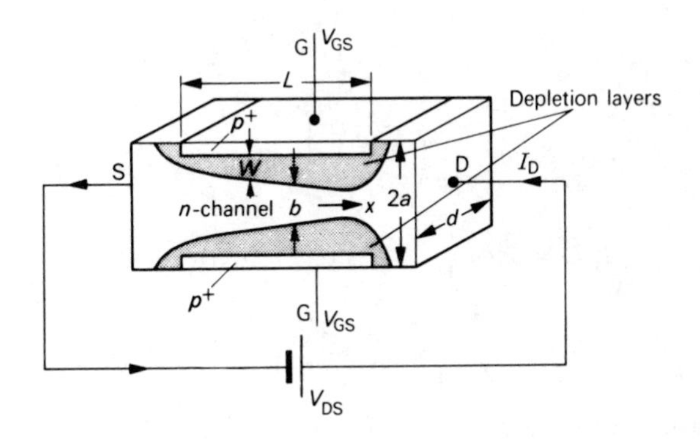

Fig. 77

thickness width w in the channel is given by

$$w = \left[\frac{2\varepsilon V}{eN_D}\right]^{1/2} \qquad (\varepsilon = \varepsilon_0\varepsilon_r)$$

where V is a function of the gate voltage V_{GS} and also of the distance x because of the flow of drain current I_D, ε is the permittivity of the material and e is the electronic charge.

At pinch-off the depletion width $w = a$ and the pinch-off voltage V_p is given by

$$V_p = \frac{eN_D a^2}{2\varepsilon}$$

and substituting for $2\varepsilon/eN_D$ from the previous equation yields

$$V = \frac{w^2}{a^2}V_p$$

or

$$w = a\left[\frac{V}{V_p}\right]^{1/2}$$

The drain current I_D which flows across the channel width b, length dx and depth d is given by

$$I_D = \sigma bd\frac{dV}{dx}$$

where σ is the conductivity of the channel. Integrating over the length of the channel from $x = 0$ with $V = V_{GS}$ to $x = L$ with $V = (V_{DS} + V_{GS})$ yields

$$\int_0^L I_D dx = \int_{V_{GS}}^{V_{DS}+V_{GS}} \sigma bd\, dV$$

or
$$I_D L = \sigma d \int_{V_{GS}}^{V_{DS}+V_{GS}} b \, dV$$

Since $b = 2a - 2w = 2a\left[1 - \left(\frac{V}{V_p}\right)^{1/2}\right]$ we obtain

$$I_D L = 2\sigma a d \int_{V_{GS}}^{V_{DS}+V_{GS}} [1 - (V/V_p)^{1/2}] \, dV$$

or
$$I_D = \frac{2\sigma a d}{L}\left[V_{DS} - \frac{2}{3}\frac{(V_{DS}+V_{GS})^{3/2}}{V_p^{1/2}} + \frac{2}{3}\frac{V_{GS}^{3/2}}{V_p^{1/2}}\right]$$

as the constant of integration is zero if $I_D = 0$ when $V_{DS} = 0$.

IGFET equation

For the induced *n*-channel device shown in Fig. 78, it is assumed that the oxide thickness is large compared to the channel depth and the channel narrows linearly over the distance *L*.

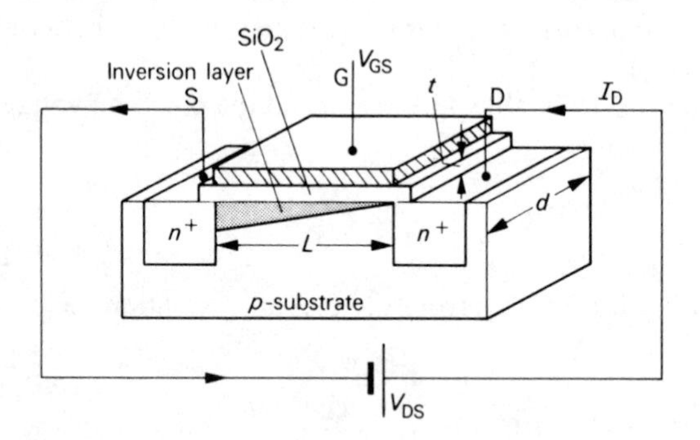

Fig. 78

If the thickness of the oxide coating is t, V_x is the voltage drop in the channel and V_p is the pinch-off voltage, the charge induced per unit area in the channel is given by

$$D = \varepsilon E = \varepsilon\left(\frac{V_{GS} - V_x - V_p}{t}\right)$$

For a channel depth d and carrier mobility μ, the drain current I_D is given by

$$I_D = \mu d D \frac{dV}{dx} = \mu d \varepsilon \left(\frac{V_{GS} - V_x - V_p}{t}\right)\frac{dV_x}{dx}$$

or
$$(V_{GS} - V_x - V_p)\,dV_x = \frac{I_D t\,dx}{\mu d\varepsilon}$$

with
$$\int_0^{V_{DS}} (V_{GS} - V_x - V_p)\,dx = \int_0^L \frac{I_D t\,dx}{\mu d\varepsilon}$$

or
$$V_{DS}(V_{GS} - V_{DS}/2 - V_p) = \frac{I_D t L}{\mu d\varepsilon}$$

with
$$I_D = \frac{\mu d\varepsilon}{tL}[V_{DS}(V_{GS} - V_{DS}/2 - V_p)]$$

as the constant on integration is zero if $I_D = 0$ when $V_{DS} = 0$.

To obtain the saturation current I_{DS} we use the relation $V_{DS} = V_{GS} - V_p$ which yields

$$I_{DS} = \frac{\mu d\varepsilon}{tL}\left[\frac{(V_{GS} - V_p)^2}{2}\right]$$

and the mutual conductance g_m in the saturated region is given by

$$g_m = \frac{\partial I_{DS}}{\partial V_{GS}} = \frac{\mu d\varepsilon}{tL}(V_{GS} - V_p)$$

and more generally we obtain

$$g_m = \frac{\partial I_D}{\partial V_{GS}} = \frac{\mu d\varepsilon V_{DS}}{tL}$$

with
$$\frac{1}{r_d} = \frac{\partial I_D}{\partial V_{DS}} = \frac{\mu d\varepsilon}{tL}[(V_{GS} - V_p) - V_{DS}]$$

Appendix F: Microwave diodes

Gunn diode [29]

In certain semiconductors such as *n*-type GaAs or GaP, oscillations can occur in bulk material rather than across a junction and this is known as the *Gunn effect*. In such devices, the current increases linearly with applied voltage up to a certain threshold value, after which it decreases with increasing voltage thus giving rise to a negative resistance effect.

The Gunn effect is due to the presence of two conduction band valleys separated by a small energy gap of about 0·36 eV as shown in Fig. 79. Energy which is gained from the applied electric field causes electrons in the lower conduction band to be transferred to the higher conduction band around fields of several kilovolts/cm. In the higher energy states, the effective mass of the electron is higher and its mobility lower. Hence, lower mobility implies lower

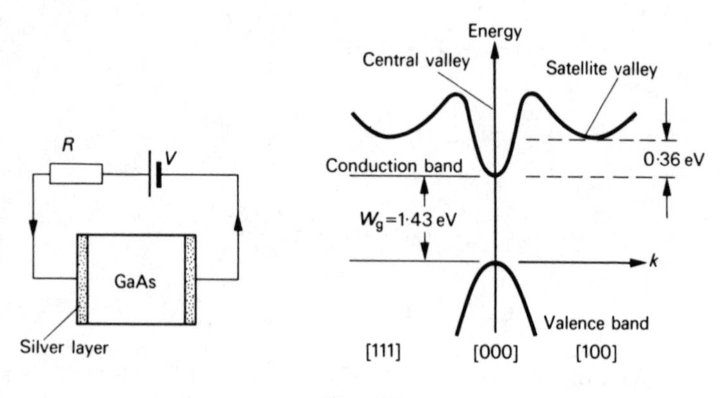

Fig. 79

current flow with increasing voltage, thus giving rise to the negative resistance effect.

It is also observed that part of the material has a high field gradient across it while the other part has a low field gradient across it. The high-field region or *domain* travels from the cathode with a drift velocity of about 10^7 m s^{-1} to the end of the material and collapses at the anode, while another domain is generated at the cathode. The sudden collapse of the domain generates a current pulse, and if the transit time through the material is such that the current pulses occur at a microwave frequency, microwave power generation is possible in an external circuit. For this to occur, the GaAs sample must be very thin (about 10 μm) and it must be placed inside a resonant cavity which may be tuned.

This is the Gunn effect mode and the oscillations are almost independent of the external circuit. However, by using the LSA mode (limited space-charge accumulation) the frequency of oscillations can be controlled to a larger extent by the external circuit. Typical frequency ranges cover the X-band (8–12 GHz) with bandwidths of about 10%. Both CW and pulsed devices are available with power levels from several milliwatts (CW) to about 100 W (pulsed), while power efficiencies are between 2–4% (CW) or 10–25% (pulsed).

Impatt diode[30]

The IMPATT diode which is also known as the 'Read diode' employs avalanche breakdown to produce charge carrier multiplication by impact ionisation. Hence, it derives its name from the words *Imp*act *A*valanche and *T*ransit *T*ime device. In one form, it consists of heavily doped p^+- and n^+-regions with n- and i-regions in between as shown in Fig. 80.

On application of a reverse bias, avalanche breakdown occurs in the p^+-n-region producing electron-hole pairs. The holes are drawn across the p^+-region

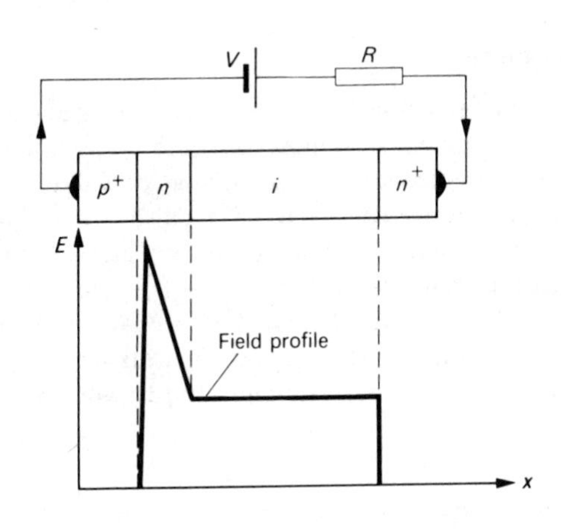

Fig. 80

to the negative terminal and can be ignored, while the electrons drift through the *n*- and *i*-regions with a constant velocity.

If an RF voltage is also present across the device (due to random noise fluctuations), the current lags the voltage by 180° due to the time lag of the avalanche effect and the transit time through the drift region, which gives rise to a negative resistance. The device can therefore function as an oscillator and, with a suitable load resistance, as an amplifier too.

Typically, output powers of 1 W (CW) and 50 W (pulsed) have been obtained at microwave frequencies. However, due to the higher voltages required for avalanche breakdown, power dissipation is high. Nevertheless, efficiencies as high as 35% have been obtained and Impatt devices are being used instead of low power klystrons in some radar applications.

Trapatt diode[31]

A variation of the Impatt diode is the TRAPATT diode which derives its name from the particular mode of operation known as *Trap*ped *P*lasma *A*valanche *T*riggered *T*ransit. In this mode of operation, space-charge suppression of the avalanche and carrier trapping play an important part in setting up the trapped plasma states. The Trapatt mode is highly efficient and operating frequencies are generally well below the transit time frequency of Impatt diodes.

To achieve this, large voltage swings must be achieved by trapping the Impatt oscillation in a high *Q* cavity. Furthermore, sufficient capacitance must be provided near the diode to maintain the high current state, and tuning or matching to the load must be possible at the Trapatt frequency. Trapatt operation has yielded output power levels of 10 to 500 W with efficiencies as high as 75% in the frequency range from 1 to 10 GHz.

Schottky-barrier diode[10]

The diode consists of a junction formed between a metal and a semiconductor which sets up a barrier potential. A metal like gold or platinum is deposited by evaporation on to a semiconductor like silicon or gallium arsenide. When forward biassed, the majority carriers, which are also termed 'hot electrons', are injected into the metal. When reverse biassed the electrons are drawn away from the junction and the diode is quickly cut-off, as minority carrier charge-storage effects are very small. Hence, the diode has rectifying properties like a *p-n* junction but its switching time is very fast. Schottky-barrier diodes are used for microwave detection and mixing, high-speed switching and parametric amplification.

Appendix G: Two-cavity and reflex klystrons

Two-cavity klystron

The output anode current can be shown to be given by

$$I = I_0 \left[1 + 2 \sum_{n=1}^{n=\infty} J_n(nk) \cos n\omega(t - l/v_0) \right]$$

where $J_n(nk)$ are the nth order Bessel functions of the first kind, l is the length of the drift space, v_0 is the initial velocity of the emerging electrons and I_0 is the steady d.c. current.

For a value of $k = 1{\cdot}5$, I when plotted reveals that it has a pulse-like waveform and is therefore rich in harmonics. The fundamental component of current at a frequency f_0 corresponding to the tuned cavity is given by

$$I_{f_0} = 2\, I_0 J_1(k) \cos \omega_0 (t - l/v_0)$$

which has a peak value $I_{f_0} = 2\, I_0 J_1(k)$. For a maximum RF output voltage V_0 (which should not exceed the d.c. beam voltage V_0), the a.c. power is $V_0 I_{f_0}/2$ and the d.c. input power is $V_0 I_0$. Hence, the efficiency η is given by

$$\eta = V_0 I_{f_0} / 2 V_0 I_0$$

which has a maximum value of 58% when $k = 1{\cdot}84$.

Reflex klystron

The frequency of oscillation of the reflex klystron depends on various parameters. Assuming a net gap field strength of E_0 and a drift length l, the deceleration of the electrons is given by

$$a = eE_0 / m$$

where e is the electronic charge and m is its mass,

If the time to travel twice the gap length is t then $t = \sqrt{2l/a}$ or

$t = \sqrt{2lm/eE_0}$. For the electron to be in the correct phase on return we must have in general

$$(n+\tfrac{3}{4})T_0 = t$$

or

$$(n+\tfrac{3}{4}) = t.f$$

where $n = 0, 1, 2$ etc. is an integer and $f = 1/T_0$ is the frequency of oscillation. Hence, we obtain

$$f = \frac{(n+\frac{3}{4})}{t} = \frac{(n+\frac{3}{4})}{\sqrt{2lm/eE_0}}$$

and the different values of n determine the various klystron *modes* which are designated as the $\frac{3}{4}$ mode ($n = 0$), $1\frac{3}{4}$ mode ($n = 1$) etc.

Appendix H: Miscellaneous topics

1 Integrated circuits[11, 32]

An integrated circuit (IC) consists of a group of electronic components such as transistors, diodes, resistors and capacitors which are all assembled on a single insulating substrate. There are mainly two types of integrated circuits which are known as the *hybrid* integrated circuit and the *monolithic* integrated circuit.

Hybrid circuits

The various active and passive components are individually mounted on the insulating surface and connected together using metallised interconnection patterns or wire-bonding techniques. The active components like transistors and diodes are made using conventional planar technology while the passive components like resistors and capacitors employ either *thin-film* or *thick-film* circuits.

In thin-film circuits which are typically about 10^{-7} m thick, the film is formed on a substrate of glass or ceramic by vacuum deposition of resistive or dielectric material. Thick-film circuits are typically about 10^{-5} m thick and the resistors and interconnections are made using silk-screen techniques.

The various active and passive components are connected externally to one another using welding or soldering techniques and a typical hybrid circuit is shown in Fig. 4. The hybrid technique is used extensively for making microwave integrated circuits.

Monolithic circuits

The various active and passive components are all mounted on the same silicon 'chip' which is typically about 1·5 mm square. There are two types of such circuits which are known as *bipolar* integrated circuits or *MOS* integrated circuits.

The bipolar circuits use conventional bipolar transistors and diodes together with passive resistors and capacitors, all of which are formed simultaneously in the same chip by planar diffusion processes. The chip is then mounted in a suitable package and connections are made to its various terminals using gold or aluminium wire-bonding, before being finally sealed.

The MOS circuits use MOS transistors, resistors and capacitors which are formed in the silicon chip using techniques similar to the bipolar circuits. The MOS circuit is simpler than the bipolar circuit as it requires only one diffusion process but its switching speed is slower. However, it is well suited to medium and large scale integration.

Monolithic circuits are usually batch produced from a silicon wafer about one or two inches in diameter. The wafer may contain several hundred integrated circuits which are separated from one another by scribing and breaking along the scribed lines. A monolithic circuit is shown in Fig. 4 and typical structural arrangements are illustrated in Fig. 81.

The widespread use of integrated circuits has led to two main applications in the electronics field. Integrated circuits are either *linear* circuits which give an output linearly related to the input, such as in ordinary amplifiers and operational amplifiers, or they are *digital* circuits which represent the discrete levels 0 or 1, such as in logic circuits used for computers.

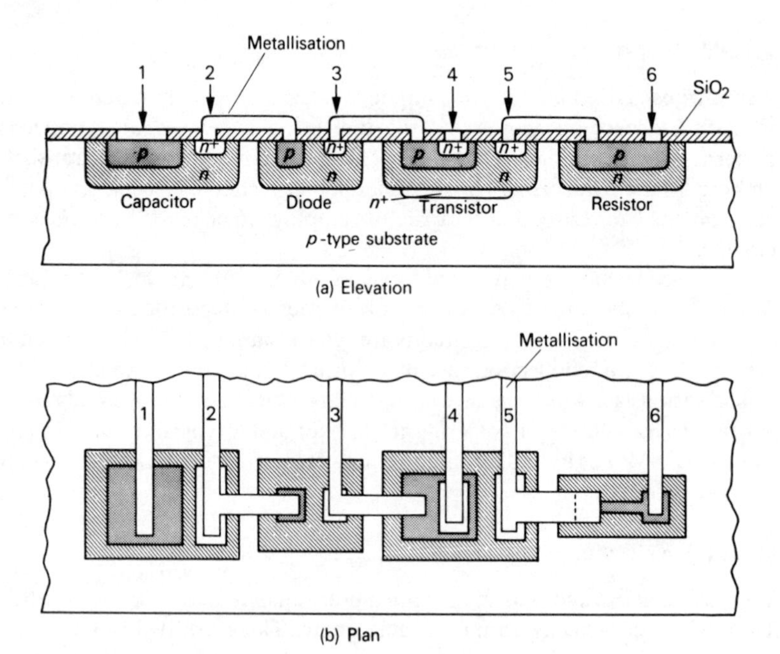

Fig. 81

Further development in this technology has inevitably led to large-scale integration(LSI) in which complete sub-systems can now be fabricated on the same silicon 'chip'. As stated earlier in the book, a typical example is the microprocessor which is the 'heart' of the microcomputer. A microprocessor chip is shown in Fig. 82.

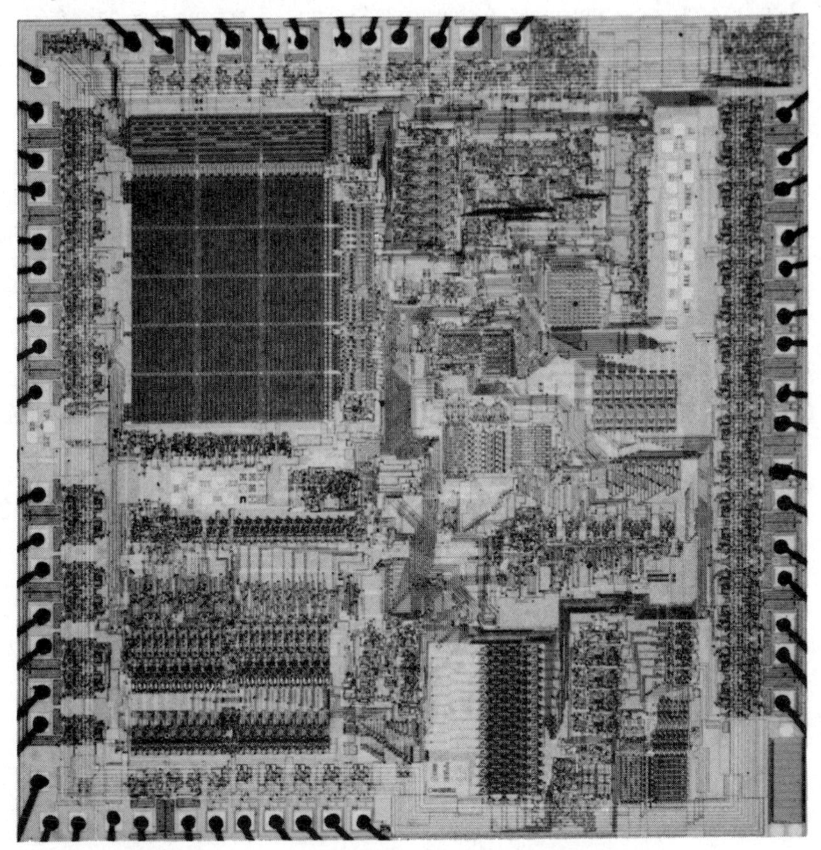

Fig. 82 16 bit microprocessor chip. (By courtesy of Texas Instruments Ltd.)

2 Charge-coupled devices[33]

The charge-coupled device (CCD) is a multiple-gate semiconductor fabricated on a silicon chip as an MOS integrated circuit. Information is stored by minority carriers as a charge packet in a potential well under a gate electrode. The charge can be moved to another gate by means of clocked pulses and so the device can operate as a shift register. The CCD can also be used for optical imaging. Light falling on the CCD structure produces electron-hole pairs proportional in number to the light intensity. The charge that collects under the gate electrodes is a charge pattern of the incident light which can be extracted as an analogue signal by clocking of the gate voltages.

The basic three-phase device shown in Fig. 83(a) consists of a linear array of electrodes arranged in groups of three phases ϕ_1, ϕ_2, ϕ_3 which are connected to their respective clocking lines. A charge packet can be introduced into the CCD by pulsing the input gate appropriately. In Fig. 83(b), a charge packet is shown under electrode ϕ_2 at time t_1 when the voltage on it is V_2. Electrodes ϕ_1 and ϕ_3 are now held at a lower voltage V_1. At time t_2, electrode ϕ_3 is pulsed with voltage V_2 and a depletion region is formed under ϕ_3 which couples to ϕ_2 because of their close proximity. If the voltages on ϕ_1 and ϕ_2 are now held at V_1, the charge transfer to electrode ϕ_3 is complete at time t_3. Charge can thus be moved from the input to the output electrodes using sequential clocking waveforms and the CCD functions as a shift register.

By using electrodes separated by two thicknesses of oxide (surface and buried electrodes), it is possible to move charge from one potential well to the next using only two clock phases. Furthermore, these *surface-channel* devices in which the charge packet is located immediately under the gate electrode can be improved upon by using the *buried-channel* CCD. In this case, the charge is moved through the bulk of the semiconductor some distance from the surface. This improves the charge transfer efficiency of the device at high frequencies.

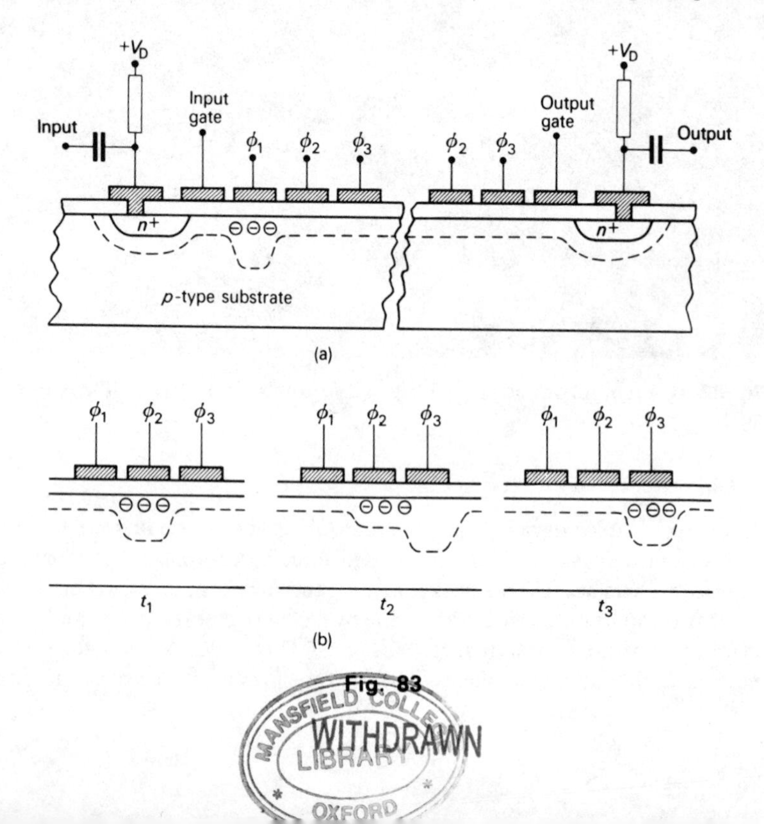

Fig. 83

Index